THE NINE EMOTIONAL LIVES OF CATS

Jeffrey Moussaieff Masson, former Sanskrit scholar and Projects Director of the Sigmund Freud Archives, has written more than a dozen books, including the bestsellers *Dogs Never Lie About Love* and *When Elephants Weep*. He lives in New Zealand with his wife, two sons and five cats.

BY THE SAME AUTHOR

The Emperor's Embrace: Fatherhood and Evolution

Dogs Never Lie About Love: Reflections and the Emotional World of Dogs

When Elephants Weep: The Emotional Lives of Animals (with Susan McCarthy)

My Father's Guru: A Journey through Spirituality and Disillusion

Final Analysis: The Making and Unmaking of a Psychoanalyst

Against Therapy: Emotional Tyranny and the Myth of Psychological Healing

Freud: The Assault on Truth

Jeffrey Moussaieff Masson

THE NINE EMOTIONAL LIVES OF CATS

A Journey into the Feline Heart

V

VINTAGE

Published by Vintage 2003

2 4 6 8 10 9 7 5 3

Copyright © Jeffrey Moussaieff Masson 2002

Jeffrey Moussaieff Masson has asserted his right under the Copyright, Designs and Patents Act, 1988 to be identified as the author of this work

This book is sold subject to the condition that it shall not by way of trade or otherwise, be lent, resold, hired out, or otherwise circulated without the publisher's prior consent in any form of binding or cover other than that in which it is published and without a similar condition including this condition being imposed on the subsequent purchaser

First published in Great Britain in 2002 by
Jonathan Cape

Vintage
Random House, 20 Vauxhall Bridge Road,
London SW1V 2SA

Random House Australia (Pty) Limited
20 Alfred Street, Milsons Point, Sydney
New South Wales 2061, Australia

Random House New Zealand Limited
18 Poland Road, Glenfield,
Auckland 10, New Zealand

Random House (Pty) Limited
Endulini, 5A Jubilee Road, Parktown 2193,
South Africa

The Random House Group Limited Reg. No. 954009
www.randomhouse.co.uk

A CIP catalogue record for this book
is available from the British Library

ISBN 0 099 44924 2

Papers used by Random House are natural, recyclable products made from wood grown in sustainable forests. The manufacturing processes conform to the environmental regulations of the country of origin

Printed and bound in Great Britain by
Mackays of Chatham plc, Chatham, Kent

For

Leila, Ilan, and Manu,

not to mention

Minna, Miki, Moko, Megalamandira, and Yossie!

Contents

Acknowledgments

I could not have written this book without the help of five cats, Minnalouche, Miki, Moko, Megalamandira, and Yossie. The book practically wrote itself; I merely watched them go about their mysterious ways, trying to capture the music of their complex emotional lives.

One human, though, made all the difference. Nancy Miller edited the first book I ever wrote. Actually, Nancy Miller edited most of the twenty books I wrote. We are together again, and I wish I could put her name on the cover. What a treasure; she must have been feline in a former life.

JEFFREY MOUSSAIEFF MASSON
Karaka Bay
Auckland, New Zealand
Summer 2002

Introduction

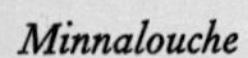

I have always loved dogs and cats, having lived with both since I was a child. Several years ago I wrote a book, *Dogs Never Lie About Love,* about the emotional lives of those wonderful animals. It was a very popular book. I noticed, however, that readers of that book would often speak to me disparagingly about cats—usually something about cats not really having an emotional life, or cats being basically indifferent. I knew this is not true, though it is untrue in ways that were not always obvious. Too many people tend to see cats as uncomplicated creatures with few emotions, at least none worth thinking about in any depth. I am convinced that, on the contrary, cats are almost pure emotion.

It is impossible to calculate the number of cat books that have been written, but there are something like five thousand currently available. Most of them are pet care manuals; of the rest, some are bad, and a few (some of which are included in my section on recommended reading) are very good indeed. Why one more? Because something is lacking in all the books I have read about cats: a serious consideration of their emotional complexity. It is not that the authors of these cat books doubt that cats have emotions, it is just that nobody has made a concerted effort to delve into their emotional world. The late Roger Caras, who was president of the ASPCA and probably America's most noted cat expert, remarked that the cat has insinuated itself into human consciousness "without revealing any of the secrets about its own feelings." This book attempts to reveal some of those secrets.

Because my wife, Leila, and I had moved from Berkeley to Boston to London over the last few years, we were unable to keep our animals with us. (My dogs are living with my friend and ex-wife, Terri, in Oakland, California; my cats of that time are living on an organic vegetable farm in Occidental, California.) I decided that when we were finally settled somewhere, I would find several cats to live with us and I would try to write about their feelings, as I had with dogs. I was not only curious about cats, I loved cats, loved watching cats, loved spending time with them, and most of all, loved thinking about cats.

The opportunity arose when we moved to New Zealand (because it is a paradise for children and we have two young children). We began building a house on a beach near Auckland, and as I looked around and saw the subtropical rain forest that surrounded the beach, the sparkling turquoise sea, the long path through what looks like a jungle down to the beach where no cars can drive, I thought: "This is an ideal place for cats." How would I find them? Of course, no sooner does one think such a thought than cats begin to appear. It is as easy to be found by a cat as it is to find one.

I went first to the SPCA, for it is always a good idea to go to an animal shelter, where cats are waiting for a home. They told me about a woman, Jane, who lived with 120 cats; she had found a particularly appealing stray. Appealing, that is, for affection. When she bent down he reached his front legs up to her and put them around her neck, pushing his head under her chin. Moreover, though large and strong, he did not fight with other cats. He was not afraid of them; he just was not aggressive. This was unusual for a stray cat, and she wondered if he had belonged to somebody. She put up notices, waited, and checked around, but nobody called for him. Maybe he had been on his own for a long time, but he was determined to belong to somebody now. He was so unusually friendly, both with people and with other cats. Would I take him? I said yes, and the very first night, Yossie (as we named him, because our five-year-old son, Ilan, was learning Hebrew and said this was a good Israeli name) slept on my chest. The curious thing is that this was a ploy, or a decision he made in his cunning little feline heart; for after a week, when he knew that he was there to stay, he stopped doing it and has never done it again. Yossie is a big—about fifteen pounds tabby, with green eyes, a long bushy tail, and striking black and brown markings on his back. He was, the vet estimated, just under two when he came to us. His life, I suspect, has not always been easy.

Like many people, I thought of cats as solitary creatures who at best tolerate other cats but are happier on their own (this is why so many people in apartments get one cat rather than two—a mistake, I now believe). Yossie, however, seemed to long for company, and not just ours. Jane took me with her one night on her rounds to feed feral cats—these are cats of whom some are stray, but most were born in the wild, though from domestic parents. Here were animals who could just as easily have wandered off to lead completely solitary lives should they have chosen to do so, yet they lived in cat colonies. The social life of these colonies was complex, if obscure to those on the outside looking in. All of these animals had been trapped, neutered, and released (the kindest thing to do with feral cats, as it ensures that no more homeless kittens will be born in their colony). Some of these cats, because they are all originally domestic (wild cats are a different species of cat altogether), would undoubtedly miss the company of friendly humans and other cats. Because they were neutered, their need for one another's company had nothing to do with propagating their genes. It must be that cats liked company. I decided, then, that Yossie needed a friend or two, or three, or four.

In theory, I strongly believe cats should be sought rather than bought, rescued rather than bred, since millions of unwanted cats are killed every year in shelters across the United States (except for the few wonderful shelters that have a no-kill policy and let such cats stay on for the rest of their natural lives). *La théorie, c'est bon* ("Theory is fine"), said the great French neurologist Jean Martin Charcot, but things happen otherwise.

I was visiting a woman who knows a great deal about cats, even if she does breed them, Twink McCabe. "Here is one cat where things went a little wrong," she said as she showed Leila, me, and Ilan, a little orange-and-white kitten. His ears were too long, his eyes too far apart, his nose too large (as if cats ever cared or even noticed such anthropocentric standards). So Miki (a name to go along with that of his friend at this cattery, Moko) came into our lives. Miki's friend Moko (the name means "tattoo" in Maori, and he looks as if he were lightly marked with a tattoo), half Burmese and half Siamese, was not a failure, and they were much attached to each other (and still are). Both were three months old. Did it not seem cruel to separate them, rather like adopting one twin from an orphanage and leaving the other to languish, bereft, alone, and brokenhearted? Or so it was put to us. I am not good at resisting the entreaties of a five-year-old or the mewing of kittens. Moko gave the appearance of a ghost-cat, with his pale white wild hair (he is a La Perm) and faint brown stripes going up his long legs. He looked more like a serval or even a lemur than a domestic cat and was as highly strung as a wild animal. *Noli me tangere* ("Don't touch me") seemed to be his motto, yet at the same time he longed to be petted.

Leila is a pediatrician. Her life is dedicated to relieving the suffering of human children in developing countries, and while she adores animals in theory, in fact she was raised without them and could easily have continued that way. She had been a gracious host to my three dogs, but it was clear she suffered no love at first sight. Moreover, she was allergic to cats—mildly so, but still allergic. However, she too succumbed to the blandishments of son, husband, and two cats. But three cats, she said, was her limit. The allergy, strangely, only got better with more exposure (self-immunization?), and soon Leila was discovering the joy of sharing a house with three felines.

I sought out people who knew about cats, and on yet another visit to a different cattery I saw my first ocicat. How repulsive to want a cat merely because of how it looked, I theorized as I stood mesmerized in front of Minnalouche (named for the cat in the beautiful poem "The Cat and the Moon" by the Irish poet W. B. Yeats), an ocicat. These cats are so called because they have the same markings as ocelots (though they are not genetically descended from them), dark brown spots and stripes edged with black on a tawny background. Our little Minna Girl, as Ilan likes to call her, is gray with black stripes and dots. She looks very much like a jungle cat, though she was tiny (only about two pounds) when we first saw her. Well, do as I say, I tell Ilan, not as I do. I justified myself to Leila by explaining that these cats were my research tools—living subjects. It seemed to me a fair exchange: they would get a good life, for as long as they lived, and I would be able to study at firsthand feline feelings.

A few weeks later, I happened to be hiking near the house of a woman who raised Bengals (*Felis bengalis*). Bengal domestic cats are only some five or six generations removed from their ancestor, the Asian leopard cat. They have a marbled look, with dark brown circles and a golden glitter. I am a total pushover for certain lines, including "I have never known such a friendly cat as this one." So Megalamandira, also about three months old, came to us (I liked the Sanskrit-seeming sound of this name that I made up). Megala is a cat, not a leopard; but his leopard blood shows sometimes. For example, he does not meow. The sound he makes is more like a birdcall, a kind of chirping, much like the Maine coon's delightful and happy "brrrp." Nor does he walk like a normal cat: he crouches and slinks along the floor, as if he were still in the jungle.

I would probably have continued like this for another dozen cats had Leila not finally put her foot down. Five was enough, she said, and I could see that she meant it. So five it would have to be, even though nine, since I have identified nine primary emotions in cats, one for each emotion, would have been ideal!

Lest it be thought that in describing four kittens and one cat, I would not be getting the response of a mature animal, it should be noted that between four and five *weeks* of age, a kitten's brain is more or less fully mature. Its senses, as well, are already as acute as those of an adult cat. This is undoubtedly because kittens need to be independent to survive at a very early age, and by four or five months most are self-sufficient. As for intelligence, consider the complex reasoning abilities needed by a kitten to negotiate his or her life. I observed this watching Miki as a very young kitten learning to sheathe his claws. When Miki first came to us at three months old, he wanted to climb my leg. I was wearing shorts. This did not bother him; he climbed my bare leg. As his claws went into my legs, I cried out in pain. He sheathed them immediately. Already by the next day he had learned, by experience, to sheathe his claws: he pawed at my legs, with no claws out, to get me to pick him up. He has never made the mistake of climbing my bare leg again. A few weeks later I was wearing blue jeans, and this time Miki did not sheathe his claws to climb onto my lap. He "knew" that the jeans were not going to be hurt. Imagine how complex the reasoning, however it is performed, must be: Miki must recognize that "I" and my clothes are not the same entity. He realizes that jeans are not a living being and cannot be hurt by claws. He knows that I am wearing something over my "self," the one that can be hurt by claws. I do not believe he sits down and reasons this out or spends time contemplating the mystery of "being" and "nonbeing." He is (thank God) no Heideggerian! Nevertheless, he knows the distinction as well as any child knows it. This is a very sophisticated philosophical ability and no small achievement.

There has been much argument about what are the "basic" emotions for humans. The list varies from three to twenty-five to seven hundred. Paul Ekman, who studies the human face across cultures, says that regardless of language, culture, and history, there are six basic facial expressions: happiness, sadness, anger, fear, disgust, and surprise. Many people who study emotion say there are seven basic emotions, since in a study of thirty-seven countries, all of these terms were found to be commonly used in different languages: anger, fear, sadness, joy, disgust, shame, and guilt. Entire books have been written about the emotions not included in this list—empathy, pity, remorse, curiosity, contempt—and whole libraries have been written about the one emotion most difficult to study scientifically (note its absence from the list): love.

If it is difficult to get scholars to agree on just what the word *emotion* means, and just how many there are in any given culture, imagine how hard it would be to get scientists (or anyone else, for that matter) to agree upon what emotions are for cats and how many they have. Many people who study the behavior of cats claim that they have nine basic senses—sight, hearing, smell, taste, touch, temperature, balance, and senses of direction and of time. I came to believe that cats have nine primary emotions—narcissism, love, contentment, attachment, jealousy, fear, anger, curiosity, and playfulness—because as I began keeping track of them, these are the ones that kept recurring. It is by no means an exhaustive list; cats can also be sad (even depressed?), affectionate, compassionate (to other cats, to people, even to dogs), disappointed (though they don't show it the way dogs do), nostalgic (you can hear it in their voices), bored, embarrassed (which they show by licking their paws in indifference), indifferent (though this may be feigned), contemplative (they are very patient in anticipating what will happen next), annoyed (at their humans for going away), confused (that you don't appreciate their offerings of headless mice and disemboweled rabbits), and just plain pleased with themselves, an emotion particularly easy to come by for cats.

What is the nature of their attachment to us, though? They are not pack animals, so—unlike dogs—they do not transfer to us feelings and loyalties meant for their own kind. It is not that we give them security; rarely does any animal threaten a cat (except dogs). They like our food but can catch their own. They do not need our warmth with their great coats. Perhaps their attachment to us can be explained in part, at least, as a kind of transference—a nostalgia for or reenactment of the time when they were kittens. Even adult cats are able, in our presence, to recapture the serene, playful joy of childhood, where we act *in loco parentis* (strictly maternal, though, since male cats have zero interest in being fathers).

We live with cats in greater intimacy than with any other animal, except dogs. I believe this has given cats access to emotions they would never have had occasion to feel in their wild state, pleasure in the company of a member of a different species, for example. We are like catnip to cats, a drug, and the addiction works both ways. People rarely merely *like* cats; they are indifferent, hate them, or adore them. For those of us who love them, the reward is great, for with no other animal is it easier and more enchanting to cross the species barrier, an almost universal desire throughout human history, than with cats.

THE NINE EMOTIONAL LIVES OF CATS

ONE

Narcissism

Moko

Very different from that faithful animal the dog, whose sentiments are all directed to the person of his master, the cat appears only to feel for himself, to live conditionally, only to partake of society that he may abuse it.

—Buffon

The frustrated woman in *The New Yorker* cartoon who asks the cat on her chair, "Am I talking to *myself*?" expects a laugh because the obvious answer is, "Yes, you are," since cats have no interest in what we say to them. But is this really so?

Many people are convinced that cats are indifferent to us. Some even go so far as to use the word *cold*, which is not really descriptive but evaluative. Most cats (of mine, only Minna Girl is a partial exception) will not come when you call them, or rather, they will come sometimes, if they feel like it, and not other times, when presumably they don't feel like it (unless there are other factors, as yet unknown to us, that decide whether a cat comes or not). This supposed indifference to humans leads some people to conclude that cats are narcissistic—in fact that narcissism is the cat's defining characteristic. Not only are cats supposed to be narcissistic, they are commonly called haughty, egotistical, egocentric, self-centered, selfish, self-absorbed, egomaniacal, smug, distant, unsociable, and aloof. As for their indifference, the phrase is usually "calculated indifference," but I doubt anyone would insist that it is calculated at all.

Narcissists lack the capacity to think about other people, to take the needs of others into consideration, to subordinate their own wishes to those of someone else. They are entirely self-involved. When I was a boy of fifteen, on an ocean liner from New York to London, I somehow struck up a friendship with a man of this description, a well-known American literary critic who was on board—the young admirer and the literary lion—and I spent much of the five days en route in his company. He spoke nonstop, always about himself, his accomplishments, his books, his admirers. It was good talk, fascinating to me at fifteen and evidently to others, for he always had a crowd. However, I knew then, though I did not know the word, that the man was a complete narcissist. He had zero interest in the ideas of anybody else around him or in anything but his own thoughts, which did indeed seem at the time more interesting than those of anyone else present. However, his fine mind could not encompass the one thought that everyone else could not avoid: He was a fool.

A cat's narcissism, if that is the word we choose to use, is not like that at all. Cats watch us all the time. Obsessively. Coldly, some would say, or at least with some detachment. They see us; they notice us. Their eyes grow big watching. They do it, some say, because they have to: we represent a superior predator, someone who might do them harm. But no, even when perfectly content, satisfied, completely out of danger, they do it. Cats take us in. We will probably never know what goes through their minds at those moments. Whatever it is, though, it is not self-absorption. The assertion, then, that cats think only about themselves is clearly wrong. Cats watch us so carefully that clearly they are thinking about *us*. But if we ask whether they think about us *in preference* to themselves, the answer is probably no.

Of course, in some sense, all animals, human or otherwise, are narcissistic to a certain degree, if narcissism can be equated with selfishness. Selfishness is built into every living creature, for none would survive without a healthy dose. Are cats more keen on survival than any other creature? It would be a strange claim. Yet cats certainly seem less altruistic than dogs, for example. I would not want to think my life depended on any of my cats. I seriously doubt that they would jeopardize their own safety to save my life. Why should they? (It seems that only dogs will risk their lives routinely, possibly because they can understand when a life is in jeopardy, whereas cats do not seem to realize this.) However, sometimes they do seem concerned. When I swim far out to sea at the beach outside my house, the four cats have a tendency to stand at the shoreline and wait for me, gazing out. Are they truly concerned, contemplating a lifesaving maneuver, or just curious? If I began to wave and shout, I doubt my cats would alter their stance.

The willingness to do something for others may be an inherited trait, common to dogs and humans but unknown to cats, having nothing to do with notions of selfishness. Why have we never heard of a service cat, like a service dog? Major economies have been driven by almost all the domesticated species—dogs (as herders and drovers), goats, sheep, pigs, cattle, water buffalo, horses. Only cats are economically insignificant—probably owing to resistance. Resistance seems an essential characteristic of cats. They resist us. Cats resist even size reduction, which we have practiced with such success on dogs. You do not find cats much bigger or smaller than other cats. They are all more or less the same size. They also resist our calls to come, to move, to obey, to present themselves, to do all the things that dogs do so easily. This drives some people crazy. Cats do not even care that it drives us crazy!

This is what some people mean when they call cats narcissistic: they will not alter their program to fit ours. It is very difficult to force a cat to do what we want. This seems to be one of the main reasons that many men in particular do not like cats: they cannot be controlled. They will not obey. Even the best-natured cat has an agenda of her own at almost all times. Even when she is doing nothing (although sleeping is hardly nothing), she does so on her own terms. Minna Girl will invariably come when I call her. Except when she will not. I call, she looks back, and then she continues on her way, not the least bit embarrassed or in any other way concerned that she has not done what I have asked. This could never happen with a dog, except under extraordinary circumstances. But for even the best-natured cat, it is an everyday experience. They hear us, they see us, they take in the request, and then they blink it away and to all appearances are completely indifferent. Yossie never does anything I ask; yet he expects me to do everything he asks. He is insistent about his food; "I want it *now*!" is his usual refrain. My cats are a lot like my five-year-old son, Ilan. "Fair's fair" is a point of view utterly beyond their grasp.

I will wake up to see one of the cats and call out, "Miki!"—hoping that he will come bounding to me, rub his nose against my face, purr madly, and in other ways proclaim his pleasure in seeing me. I love morning greetings—two beings demonstrating the joy they feel in seeing one another again after a period of separation. But Miki walks past me without even pausing. All of the cats do this at some point or other. They act as if I were not there, as if I hardly mattered in their lives. Later in the same day, they will be running and playing with me on the beach, their eyes shining with pleasure, clearly delighted we are all together. I am learning to leave my expectations behind and take what comes as it comes. I seem to have no choice with cats.

Is what looks to us like studied indifference really that, or is it just that we do not entirely understand cat rules of behavior? Cats might assume, for example, that we can read their minds: "Can't you see I am thinking of something else entirely?"—in which case for us to insist on our own agenda would be impolite from the cat's point of view. They have something they are intent upon, a place they would rather be, a task they would rather perform, and our insistence that they conform to our plan is simply irrelevant to them. It does not occur to them to obey any request they do not themselves wish to perform or that is not self-generated.

In all probability, this comes for the most part from their ancestry, from living as solitary hunters. The direct ancestor of our domestic cat is the African wildcat. Very little is known about the behavior of these creatures, because it is so difficult to approach them, but as far as we know, they are entirely solitary. Of course, cats (even apart from lions and cheetahs, which live in social groups) are not nearly as solitary as we make them out to be, but compared to the social canids—dogs, jackals, even most foxes—they certainly are. They spend most of their lives, and even our domestic cats spend a good part of every day, entirely on their own. Therefore they have come to depend on themselves, and on their own judgment, without the need to refer to a higher authority or even a group opinion.

It would be hard for humans to alter such a fundamental mind-set, deeply engraved as it is in the cat's genes. For ten million years, cats were completely solitary creatures, interacting only for their first ten weeks with their mother and with other cats as sexual partners from time to time. It is only relatively recently in their history—four thousand years ago—that they were domesticated. And even though they have lived with humans for four thousand years, it is not clear that cats are completely domesticated in the same way that dogs are. Cats do not look to us for advice, supervision, or training. We improve their environment; above all, we make it safer and more reliable, but we do not control it entirely. No cats ever completely give up the sense that they must make their way alone in a hostile world. I say this because cats, perhaps the only domesticated animal that can do this, so easily revert to a wild state, what we call feral living, as if they had been prepared for it all their lives. A feral cat is not a "wild" cat in the literal sense of the term. A feral cat is one born outside of a regular home, whose mother or a more distant relation was at some point a domestic cat, probably abandoned. Farm cats, many of whom are already half feral, and even the most pampered apartment cats could probably become feral without any great challenge to their sense of self. Some animal behaviorists go so far as to say that the cat is not entirely domesticated to begin with. Cats do not make a pact with us, as do dogs, such that our presence can compensate for the loss of mastery of their external environment.

Part of the reason we see cats as aloof, narcissistic, distant, or whatever other term we use is that we contrast them with dogs, who often appear to reflect on their own inadequacy. I am thinking in particular of the unusual trait dogs have of expressing guilt (or at least shame—it is often difficult to determine the difference) when they have done something wrong in our absence. We return home and are greeted by an embarrassed dog at the door, one eager to let us know that he has done something we are not going to like. The dog knows what he has done and is remorseful, or so it seems to us. He may be anticipating punishment, rather than feeling true guilt for his actions, but whatever the source of the feeling, the feeling itself is obvious. This is a remarkable capacity and one humans admire greatly, no doubt because we share it.

The cat has no such capacity. Cats do plenty of things in our absence that we do not like, but I have never seen a cat ask for forgiveness. Moko rolls in the wet sand on the beach, then returns to the house and rolls on the new purple couch. Minna Girl brings in oversize banana leaves, her idea of hunting, and strews them over our living room floor. Yossie finds household sponges, which he places in strategic spots around the kitchen. Miki collects butterflies. When we find what they have done and blame them, they are clearly ill at ease. It makes them uncomfortable to see us angry (so much so that they will often walk away). But guilt? Remorse? Forget about it! Guilt is just not part of their emotional repertoire. Has anyone ever called a cat sheepish? I don't think so. Is this a defect? Of course not. It is an artifact of their evolutionary history. When dogs who live in packs do something wrong, the whole pack suffers. Reprimand must be swift and automatic. When a solitary hunter such as the cat makes a mistake, nobody suffers but the cat himself. Nor is there anyone else to complain. It is just not part of the cat's mental makeup, then, to feel guilty.

If cats do not feel sorry when they cause us unhappiness, it does not mean they cannot feel sorry about our misfortunes. Almost all of us who have lived with cats have had the experience of a cat knowing we are sad and attempting to cheer us up. When I had a high fever, one of my cats sat on my chest, purring, for days on end. A cynical vet friend told me it was because it was the warmest spot in the house, but I like to think my cat wanted to make me feel better (he did). I have heard of cats licking away tears. Moreover, there are many reports of cats mourning the death or absence of a close friend.

Nobody is entirely certain whether cats feel shame, the

watered-down version of guilt. Shame implies that you are being watched, or even that you imagine you are being watched, and you are ashamed of what you have done in the eyes of another. Guilt, on the other hand, is something you feel quite apart from the prospect of how others view what you have done. It is you yourself judging your behavior by your own inner standards, not by an external yardstick. Cats rarely, I would be prepared to bet, spend sleepless nights tortured over whether we disapprove of their previous day's behavior.

Yossie was a brute the other day. Ilan was stroking him ever so gently, when for no apparent reason Yossie struck out at him and scratched him on his face. It was more than a warning tap, and it came without any warning in any event. Ilan was mortified and hurt. He cried loudly and indignantly: "I was just petting him!" True. There was no discernible motive, nothing to warrant the attack. We all yelled at Yossie, who left the room slowly but with no obvious concern. Did he regret what he had done at any moment that day? I frankly do not think he gave it so much as a passing thought. It is puzzling, because we had made it obvious that we were not at all happy at his behavior. He might have fretted that this time he had gone too far, that maybe we would retaliate and take away some favorite pastime (like eating all day long). However, had we persisted in our anger, it is he who would have been puzzled. Why all the fuss? Moreover, here is the worst part of all this: Had he hurt Ilan badly, Yossie still would not have regretted what he had done.

This is what I mean by saying that regret or guilt are completely absent in cats. The feeling just is not there. Not only has it not been hardwired into a cat, you cannot put it there, either, no matter how much effort you make. Cats were never meant to feel guilty. If Yossie goes too far again one day, and we have to send him off to live elsewhere, I can be sure he will never understand the connection to his own behavior. He might be angry, but he will never be angry with himself. Nor does there appear to be any way to enlighten him. He will never suffer the "aha" experience of "Oh my God, it's me, I am my own worst enemy!"

Is this lack of regret, shame, and guilt proof of narcissism? Only if we decide in advance that narcissism is the only way to explain the inability to feel these things. Cats, however, speaking from an evolutionary point of view, were never meant to experience many of the emotions that are "other based"—that is, that depend on the presence of another, significant being. Species like humans and dogs that are sociable and depend on others for security, safety, or pleasure need signals to show they are sorry; they have to be able to apologize. A solitary hunter has no need for such self-debasement, such a loss of face. The human presence has modified cats, but only to a degree. Cats do not apologize. I am unaware of any research on the concept of "sorry" in species other than dogs and primates. There could well be subtle signs, in other species, however, that we simply have not noticed yet.

Cats seem very aware of their dignity and go to great lengths to maintain it. A woman wrote to me about her Siamese cat who loved to play but seemed to feel it was undignified. She hated people to actually see her playing. If she thought somebody had spied her, she would stop immediately, sit regally upright, and behave as if she had been in that position for hours.

I watch with fascination as the older, wiser Yossie puts up with the antics of the two kittens with something that looks very much like dignity. When he has had enough, he slowly raises a single paw in warning, and because they are kittens and understand so little beyond their own needs, they ignore it, until he brings it down closer to them, at first tentatively and then slowly, with more force. He is losing patience. His dignity is beginning to fray. The term is not totally meaningless: he does not hit at them with intemperate anger. He is not furious, just annoyed. It is mild. He knows his superiority. He just wants his peace. He does not move crazily as they do. He has his dignity.

How unseemly, how humiliating, how mortifying, it would be for such a supremely self-confident creature to falter. Mortification, the loss of dignity, seems like such a human trait that it fascinates people when they observe it, or believe to have observed it, in cats. I have seen cats pretending not to be mortified when they are. When you scold a cat, she has a tendency to begin washing herself, paying zero attention, quite on purpose. If she leaps and misses a ledge, for example, she also starts to wash herself, as if it were of no consequence. Humans, when they fail, feel humiliated. It is worse if other people witnessed our humiliation. Then we feel deeply mortified. Is this the same feeling cats have? It is hard to say whether the need to maintain dignity is a form of self-absorption or if the self-absorption is merely a ploy to maintain dignity. The origin, though, at least in cats and possibly in humans, too, is an old survival mechanism: not to show another predator (us, in this case) that they are weak. This is less a feeling than a strategy.

Their need to save face colors their ability to appreciate a joke with you, even at you, but seldom one directed at themselves. I remember being in the living room when Minnalouche leapt at a fly, missed, and fell off the table. When we laughed, the cat ostentatiously turned her back on the room and examined her paws. This only increased the laughter, at which point she walked slowly out of the room and would not return when called. It is hard to believe she did not feel humiliated. Cats do not like being laughed at (few animals do; research has shown that elephants, the great apes, and dogs all seem to be able to interpret such human laughter as hostile, not humorous).

Some cats do seem to like to be looked at, though. That consummate cat observer Doris Lessing, in her classic book *Particularly Cats*, writes of an especially beautiful cat, "She was as arrogantly aware of herself as a pretty girl." Many cat lovers would insist that their cats know perfectly well when they look their loveliest (in repose) or when they are well displayed; they seem to enjoy the human admiration that results. Nevertheless, applying human aesthetics to cats can be risky. Does Minnalouche know how beautiful she is—compared to other cats, I mean? Are such distinctions as we make ever part of the cat's world, as in "I am prettier than Yossie, but Yossie is prettier than Miki"? This seems doubtful. It is humans who find Minna's gray fur, with the spots like an ocelot's, attractive. Cats do not hold cat shows, humans hold them, although when I have asked some of the people who display there whether they think their cats are aware of the stakes or understand the game at all, many say they do, that the cats enjoy it and are competitive about their own looks. I do not think so. This is, most people would agree, an example of humans projecting their own values onto their cats.

Many people feel that cats are not susceptible to gratitude. Dog lovers know that dogs display gratitude in an obvious way, jumping up and more or less shouting, "Thank you thank you thank you!" Cats are usually more subtle, but not always, as Doris Lessing demonstrates in a passage where she writes about Rufus, a stray she was determined to keep out of her household with its resident cats. She lost the battle, of course, and Rufus's purr rumbled through the house: "He wanted us to know he was grateful. It was a calculated purr." He was saying: "Look, I am grateful, and I am telling you so." Nonetheless, he was confined to the kitchen. Then one day he made a dramatic apparition into the sitting room: "Here was the embodiment of the dispossessed, the insulted, the injured, making himself felt by the warm, the fed, the privileged." When he triumphed and was let in, "he knew he was lucky and wanted us to know he understood the value of what he was getting." Hence the calculated purring. Like anything that anyone writes about another species, this is merely an interpretation, and if I find it a convincing portrait, perhaps this is because it is so well described. However, it also corresponds to what I have seen with my five cats: sometimes when I offer them food they especially like, before they even eat it, they come to me and rub themselves against my leg and purr loudly. Can this be anything except "Thank you"?

Cats appear to be always independent and self-sufficient. It is a "hubris that has created a very dangerous fiction," according to Roger Caras. In his book *A Cat Is Watching*, he says that this belief has led people to ignore the real dangers in the lives of cats. He writes, engaging in a bit of charming anthropomorphism, that "the proud little hairballs, in fact, would be far better off if they would just admit their vulnerability and throw themselves on the mercy of the merciful among us. They are too proud for their own good. A little less strutting might lengthen feline life expectancy." This implies that cats deliberately assume airs. I do not think cats do this at all. Cats are not too proud to ask for help. It is one of their endearing traits that when they get themselves into trouble and believe we can help, they immediately turn to us for extrication. I help my cats on a daily basis, and they not only ask for it, but also acknowledge it after it happens. To acknowledge dependence and to be grateful for small favors is not a characteristic of a supreme egoist or a narcissist.

When Moko was stuck in a tree and I helped him down, he purred with gratitude. He did not pretend that he did not require my help; on the contrary, his eyes were glowing with pleasure when I appeared, and he chirped as soon as he heard my voice. When he was out of the tree, he rolled around in ecstasy, filled with the thankfulness of the saved. He had no pride. It is not that cats say to us, "I do not need your help," but rather, as self-sufficient creatures used to being forced to rely on themselves, they are not always expecting help. They evolved to depend on themselves, and it is something of a miracle that they are learning, in a very short period (for one does not expect biological/genetic change in the mere four thousand years that cats have shared their lives with us), that we can be, if not entirely relied on, nonetheless a good bet.

• • •

Narcissistic personality disorder is the most popular mental diagnosis of our time. A standard psychiatric dictionary (Hinsie & Campbell) defines it as the "hypercathexis of the self and/or a hypocathexis of objects in the environment." (Psychiatrists have a wonderful way with words.) This means only an overpreoccupation with oneself and an underpreoccupation with others. We call this a disease in humans only because most people do not suffer from it. If they did, it would be the norm and not a matter for psychological concern. Never mind whether we wish to dispute this on philosophical grounds; the point is that cats are not behaving *out* of the norm when they appear to us to be narcissistic. On the contrary, they are doing what cats are expected to do—in fact, what cats evolved to do. As I have stressed earlier and will stress again in this book, because it is the key to the cats' mind: They did not evolve as pack animals, as sociable beings, who need to think about the safety and well-being of their friends and relatives before their own. Cats are anything but a hive of bees.

If we cannot call cats narcissistic, I think it is nevertheless fair to say that we cannot speak of feline humility, either. Rarely—in fact, never—have I heard anybody refer to a cat as humble. Humility implies a kind of recognition and acceptance of one's own unimportance. Humans appreciate this emotion. It is not so much a feeling as an attitude, and sometimes we wonder whether it is not a bit of a put-on. Not in a dog, of course, but in us. Psychological fashions change: there was a period when we highly valued humility in a person. Today we are more likely to diagnose a lack of self-esteem. Humility is not a characteristic we associate with cats. Why is this? Not because it is exclusive to humans. I have heard about, and have lived with, humble dogs—that is, dogs who make few demands, who seem to think of their place in the house as south, way below that of the resident cat. If a cat were to behave like that, we would think he or she was sick or suffering in some way. The reason is not far to seek: As pack animals, some dogs are on top and others are on the bottom.

My father was fond of a phrase from Blake's "Auguries of Innocence" that drove the rest of us crazy, because he used it as a cliché to explain every inequality in the world:

> Some are Born to sweet Delight.
> Some are Born to Endless Night.

This simple catchall philosophy allowed him to reconcile himself to his own good fortune and the sufferings of others (which, as La Rochefoucauld said, we seem to have an infinite capacity to bear). Dogs, it would appear, accept this rather primitive philosophy, but you would find it impossible to convince any cat that it was accurate. They are all born to sheer delight. However, nobody blames a cat for lacking humility, except perhaps people who do not like cats. Once again, the explanation is to be found in their evolutionary history. As a solitary hunter, what value could there be in feline modesty? Cats do not benefit from putting themselves down.

This does not mean, though, that cats are smug, even if they exhibit what looks to us like smugness, a kind of arrogant superiority. "Smugness" is rarely used of a person as a tribute. However, in the case of cats, here is an attitude that is almost universally ascribed to them, and not only by people who do not like cats, but by cat lovers, too. They mean it as a compliment, somewhat like the obverse side of dignity. Nevertheless, smugness implies that one is smug vis-à-vis someone else; it is not an emotion that can be felt on its own, in the absence of an object. It is the satisfaction to be derived from a sense of superiority, of being brighter, or more talented, or more capable, than somebody else. Of course cats do not experience smugness in this sense.

Are cats narcissistic? No. They have been called narcissists only because we use the words loosely. Our use of the term is, well, narcissistic. We think only about people, not about other sentient creatures who share so much of our biology and our emotional lives and yet are different from us. The fact that people use the words *narcissistic*, *selfish*, and *self-centered* when talking about cats only shows that we tend to become what we discuss. Cats are not minipeople; they are profoundly other. Cats have their own versions of our most common emotions.

They are not identical to ours. In the strictly psychiatric use of the term, a narcissist has a problem, whether he or she sees it as such or not. Cats do not have a problem. In the popular use of the term *narcissist*, to refer to somebody who thinks only of him- or herself, cats think a good deal about others, including us. What they think of us, well, that is a different question.

Possibly the best refutation of the narcissism attributed to cats is their ability to love. How can we doubt that ability? I hear from just about everyone who lives with cats how hard it is on them to be separated from their human companions. One woman told me that when she returned from a three-week absence, her cat looked at her and began to "speak"—meowing, howling—something he did only rarely, and he did not stop for a solid hour. Clearly, he had things on his mind that he needed to communicate. What was he saying? She interpreted it as something equivalent to: "Where have you been? How could you have left me for so long? I missed you, I missed you, I missed you!" Another tells me that when she went away for a few days, her neighbor who feeds the cat was awakened at two A.M. by the loudest, most baleful wailing. It sounded, she said, like a baby who had been abandoned.

Perhaps we have a tendency to interpret as smugness, or indifference, or superiority, or superciliousness, or disdain, or aloofness, or whatever other negative word we can think of, what is really a form of self-satisfaction—that is, that the cat is happy to be himself. Cats do not need constant reassurance from us. They already know how fine they are. Why repeat the obvious? Because we are not used to this feeling, in ourselves or in our other animal companions, we do not readily recognize it and have no vocabulary for it. Cats are contented with their lot, with or without our approval. They have less need of us than we would like. It would flatter us to think that cats cannot survive without us, that they need us emotionally as well as materially, whereas in fact they probably do not need us in either sphere. This wounds our vanity. We need cats to need us. It unnerves us that they do not. However, if they do not need us, they nonetheless seem to love us.

TWO

Love

Yossie

To most of us, the most beautiful word in any language is our own name spoken to us with love by somebody we love back. Cats are no exception. It is strange, when you think about it, that cats love to hear us repeat their name, but they do. Cats know their name of course; they know that the name we speak is only for them—their very own, singular name. They know, too, that when we say it with love, as we often do, we are saying something special to them. This gives them enormous pleasure. What is the nature of that pleasure? Why would these solitary animals be driven to ecstasy by the sound of their own name? What are they hearing?

If you ask people who live with cats (I do this all the time) what their cat likes best about them, the first response is invariably something to do with material life. I asked the novelist Stephanie Johnson what her cat loved in her, and she said, "I know how to open the fridge door, and all he ever does, I swear, is stare at that door. If you were to look into his mind, you would see a white oblong shape, strangely resembling our family fridge." If a cat's life is reduced to sleeping and eating, then clearly the cat will think about this all the time. However, a cat may love you for many different things, some of them material, some of them intangible, some of them obvious, some obscure. We should not expect to exhaust the list any time soon. Even if we could ask them, they might not be able to enumerate all the ways they love us.

Nobody who lives with cats would believe they are indifferent to humans. Nevertheless, some hard-nosed scientists are not convinced that cats feel love and affection. In the nineteenth century, Nathaniel Southgate Shaler, dean of science at Harvard, a great foe of the cat and lover of evidence, said, "I have been unable to find any authenticated instances which go to show the existence in cats of any real love for their masters." We might argue endlessly about what exactly an authenticated instance would be, but of cat love for humans there are no end to stories, testimonies, and direct observation. Among the best is an account by Frances and Richard Lockridge, mystery story writers (their popular North tales first appeared in *The New Yorker* in 1936), in their invaluable book, *Cats and People* (first published in 1950 and still one of the best books ever written about cats), where they write about their beloved Pammy. When Richard was absent during part of World War II, "Pammy was brokenhearted—one does not like to use terms so extreme, but other terms are inadequate. The bottom dropped out of Pammy's life." What is the evidence?

She would go to the door of the small room where Richard

always was. She knew that he was not there, and after looking into the room, she would "raise her head and give a small, hopeless cry." She would turn away and wander the apartment restlessly, returning again and again to the room, only to find it empty. If she heard the front door open, she was all ears, but she would know instantly from the footsteps that it was the wrong sound, and she would cry and wander again.

The Lockridges write, "But if, during those weeks, she did not feel deeply the loss of someone she loved, then the actions of cats and men make no sense at all, and the words we use have no meaning." As I was writing these lines, Minnalouche was sitting perched on the top of my computer monitor. She looked down at my fingers racing across the keyboard, following their movement. Then she glanced up at my face and blinked. I blinked back, and that gesture alone (one of friendship in cat language, indicating that one has no predatory intent) so delighted her that she began to purr loudly. I had not petted her; it was the mere *idea* of my friendship that had pleased her so. The Lockridges write of their Siamese cat, Martini, "When she blinks a little, gently, as we speak her name, it would be easy to think that there is a kind of adoration in her mind."

Because cats lack a protective antibacterial enzyme (lysozyme—humans and most other animals have it) in their tears, they can blink as infrequently as once in five minutes. (It is hard to believe that blinking exposes the eyes of domestic cats to more bacteria than a wild cat, who would not have to blink in a friendly matter except under unusual circumstances, but perhaps it is so, one of the disadvantages of domestication for cats.) If cats blink rarely, do they do so only when they want to suggest friendly feelings? The medieval church fathers thought it was evil for a cat to stare at you and not blink. Humans are more like the social dog in this respect and respond to a stare with aggression. My cats stare out of affection. They also signal their lack of hostility by deliberately blinking from time to time, perhaps afraid I will misunderstand. Megala is a frequent blinker. He has also developed a blocked tear duct about which little can be done medically. Are the two connected?

Cats look away or blink when feeling friendly. Dogs, under similar conditions, close their eyes slightly. Humans, according to James Serpell, director of thr Center for the Interaction of Animals and Society, at the University of Pennsylvania, are probably the only mammal to use eye-to-eye contact as a means of expressing intimacy (in monkeys, dogs, and cats, it conveys hostility). This probably goes back to the nursing dyad, where the infant gazes, with something approaching love, into the eyes of the mother. When petted, cats will look up at their companions with a look we interpret—correctly, I believe—as adoration.

• • •

Like most cats, mine do not like rain. Yet when Leila, Ilan, Manu, and I walk up the hill in the evening in the rain, such is the cats' devotion that they come with us. I could not quite credit it the first time it happened, but they do it often enough now to have convinced me that their own comfort is secondary to them; first comes the desire to be with us. The love could be for us, for the adventure, for the variety, but it is beyond question that the cats are doing something because they *want to*, because they derive pleasure from it. Moreover, since they walk in the rain only when we are there, never alone, the love of our companionship must be a strong component of their pleasure.

I believe not only that cats think about us on a fairly regular basis, but also that they can read our intentions or, deeper still, our true feelings about them. A woman I know, mad about cats, took one look at Yossie and suffered a *coup de foudre*. She got down on all fours and cooed to him. Yossie looked stunned: ran over, rubbed noses, then began the most shameless display of requited love—"Yes, oh yes, me too, I feel just as you do, I am yours forever." It was unmistakable. Now, Yossie clearly likes me; he does more than tolerate me, but he is not wild about me. He was instantly wild about her. Why? Because he read her so clearly. Her love for him was unconditional, or so it must have appeared to Yossie. Maybe he sees in me some hesitation, some uncertainty about his "character," about how trustworthy he is with children, which does indeed make me hold back total devotion to him.

It still seems to me, though, that the cats like one another more than they like me; they never groom me as they do each other. Today I brought Moko back from the vet, where he was treated for ear mites, and Miki cleaned out his ear with his tongue while Moko purred. If cats have no other cat, then they tend to treat us as cats and will groom us, but given a choice, they prefer their own kind. Not surprising, really. The surprise is that they like us as much as they do, especially since cats rarely like other animals at all, except for dogs. Cats can be very close to dogs, so close that they become inseparable companions, and many people have told me that when the dog dies, the cat goes into mourning. When I had dogs and cats together, they would often set off on adventures with each other. Cats can like a rabbit or even a mouse, but they will rarely bond in the same deep way with those animals as they do with us and sometimes with dogs. What is it that cats, humans, and dogs have in common? Our devotion to love.

What about cats who seem to like no other cat? There are cats who resent the intrusion of any new cat into their home (which has become their territory), they disapprove of affection shown a strange and needy animal, and they want all the food, all the attention, and all the love for themselves. Many people have only one cat in their home because they have been convinced, by experience or by mythology, that cats prefer it that way. I do not believe it is true (though I wonder why my cats sniff at me so intensely when I have been out cavorting with strange cats). My five cats are not related and came to us at different times, yet four of the five are good friends. They sleep together, eat together, play together, and take walks together. Here is the key: If cats are living in stressful conditions, where there is not enough food or not enough stimulation, they may well resent the presence of other cats who take away these vital requirements. Even feral cats act quite differently according to whether there is plenty of food or whether they are really on their own. When there is plenty of food, the cats tend to not only be tolerant of one another, but actually form friendships. I would not say that most domestic cats are loners, only that they evolved to be loners and can easily revert to it.

In a short story, "*Cats,*" by the writer Robley Wilson Jr., a woman explains to her new beau, a therapist, that she has two cats who "think they are brothers. It can't do any harm if I let them go on thinking so, can it?" He responds stiffly, "It's all right to be silly over animals, so long as you know you're being silly." Bad answer. He is soon history, and at the end of the story, the woman watches her cats moving down her long driveway and she suddenly feels "the welling-up of emotions she hadn't known since the days when her sons went off to school together. *That's love*, she told herself. *There's nothing foolish about love.*" It is a wonderful phrase, that there is nothing silly about love, whether expressed for one's children or for the close animal companions in one's life.

We definitely feel love for cats, and by extension, this probably tells us something about their capacity to love us back. You do not find people feeling this kind of love, for example, for insects, even though insects fascinate many people. We would find it odd if somebody said they were mad about their praying mantis, in fact, in love with it. That is because we know (or think we know) for certain that a praying mantis cannot love you back.

Megalamandira has a way of greeting me after a long (several hours) absence: He leaps into my lap when I am sitting at the computer and throws his head hard up against my chin, then stretches his front legs over the top of my head. He is clearly exuberant when he does it and is careful to keep his claws sheathed. It is not like the greeting of any other cat. This is because Megala has leopard blood in him, and this is how leopards greet one another. You may not wish to call that gesture evidence of love, but neither can you call it indifference. I concede that we may not have the right word for what the cat is feeling. We have only human words, and these are restricted. For most of human history, after all, they have been used only for humans, not for other animals. Our vocabulary is still poor, but it is growing, and there may come a time when we find words that approximate more closely to what the cat feels. We will certainly never be in the position where the cat nods in agreement, and says, "Yes, that is exactly what I meant." Yet we can be fairly certain that increasingly careful observation will show that cats feel positive emotions in ways that are familiar to humans. Refusing to acknowledge the similarity is a perversion of strict science, as if to do so would somehow diminish our own humanity. On the contrary, to recognize the parallel between humans and cats is not only fascinating, it enhances our own emotional complexity.

Darwin, at the beginning of his wonderful book *The Expression of the Emotions in Man and Animals*, discussed what he called the "principle of antithesis." He describes how a cat who is feeling affectionate and caressing her "master" takes on the attitude that is the exact antithesis of the same cat who is about to attack an enemy. The attitude he describes is the opposite in almost every respect: "She now stands upright with her back slightly arched, which makes the hair appear rather rough, but it does not bristle; her tail, instead of being extended and lashed from side to side, is held quite stiff and perpendicularly upwards; her ears are erect and pointed; her mouth is closed; and she rubs against her master with a purr instead of a growl."

Unlike Darwin, I believe that emotions guide just about everything a cat does, including its antithetical gestures. Therefore, I want to take over the name and suggest my own principle of antithesis, applying it strictly to something that cats feel. As we all know, cats are very careful and proprietary about their extraordinary physical equipment. A cat's paws, whiskers, nails, ears, and belly are all ultrasensitive, perhaps because they could so easily be damaged, especially by a well-meaning but ignorant human.

My principle of antithesis says that it is sometimes these very body parts that our cat will present to us. It seems to be a way to display love, by showing trust in the highest degree. "Here is where I am most vulnerable, where I can most easily be harmed. Touch me here. Play with my nails. Pull my whiskers, but gently; scratch my ears and belly." What an amazing gesture. Many cats will never make it (Moko); others can think of nothing more wonderful. Does it amuse them? Give them a pleasant fright? Is it like a human counterphobia (climbing mountains when you are frightened of heights)? Alternatively, is it merely a way for them to demonstrate their intimacy and affection in a manner that we find persuasive? Is it that they know, and we know, and they know that we know, that only love could make them trust us to this degree? Whatever it is, clearly the *feeling* of trust, or security, that accompanies the act is paramount. It is as if they were proving to themselves (and us?) that they can overcome an instinct in favor of a friend, and this ability gives them pleasure.

Something new that the cats have only recently started doing with me seems part and parcel of this emotional antithesis:

As we walk down the path to the house (never up the path to the road, for some reason), the cats will bound ahead of me and then suddenly throw themselves at my feet, stretching full out. It is all I can do to stop myself from tripping over them. I have to jump to avoid them. That seems to be what they want to achieve. It appears to amuse them. At the same time, they are very exposed, for should I misstep, I would hurt them. In this case, it could be a sense of humor they are demonstrating. They are daring me to hurt them because they know I will not—making themselves vulnerable because they trust me.

Miki has a variation on this. When I take the wheelbarrow up the hill, he will hurl himself in front of the wheel, forcing me to come to an abrupt halt in order not to run him over. It is clearly a game for him, but one in which I prove myself, over and over, to be trustworthy. I wonder what would happen if I were to accidentally run him over? Would he know I had made a mistake, or would he never play that game again or, worse, never trust me again? Cats do not give us too many chances. Abuse a cat's trust twice and you could be history, for cats are much less forgiving than dogs and will often lose their trust in you should you behave badly. A dog gives you infinite slack; not so a cat.

Do cats demonstrate this same trust with other cats? I have never seen it, and in this sense, I believe they have more trust in humans than in other cats. Trust is nevertheless rare in cats. The contrast with dogs is striking: if we ask a dog to do something dangerous (jump into the surf, for example), he or she will trust us and do it. You can rarely request anything of a cat, let alone something that would actually be dangerous. I have to give Moko his ear mite medication once a week. It is not painful, just annoying, but when he sees me coming with the bottle he races away. "Stay" is an absurd request to any cat, whereas if we said this to a dog, he or she would comply. This absence of trust—or obedience—in cats is not, however, proof of an absence of love. It is just one more example of how cats evolved to be self-sufficient. Given this evolution, their undeniable affection for us is all the more to be appreciated.

How far will a cat take love? Do cats mourn? Can they die from love? My sister Linda was visiting her vet some years back, when she saw a cat with all four of his paws heavily bandaged. "What happened?" she inquired. The vet said that the cat's owner had thrown himself out of a ten-story window the day before and died. The cat had followed him; she was badly injured but lived. How can we explain this extraordinary behavior? Could the cat have known the consequences of the jump? It is hard to believe that she did, but even harder to believe that she did not. Cats, with their arboreal ancestry, need to know exactly where they are all the time in space—otherwise a fall from a tall tree could prove fatal. In fact, when cats fall from high places they usually do survive. In a 1987 study from the *Journal of the American Veterinary Medical Association*, two vets examined 132 cases of cats who had fallen out of high-rise windows and were brought to the Animal Medical Center in New York for treatment. On average, the cats fell 5.5 stories, yet 90 percent survived, although many had injuries. One cat fell more than twenty stories without serious injury. The mechanism is simple: They spread out all four legs, to give their bodies a parachute effect, and as they approach the ground, they break the fall by landing on their front feet first and bending their spine to absorb most of the shock. This cat who jumped out of the window was literally "following" her beloved human friend, even if it meant risking her life.

Some people are skeptical of the ability of a cat to commit suicide, perhaps because they do not consider cats capable of the love for another that would override their fabled self-love. Many vets say they have never heard of an authenticated case. Susan Little, a veterinarian at the Baytown Cat Hospital in Ottawa and a specialist in feline medicine, writes to me that in twenty years of practice devoted primarily to cats, "I have never seen or heard of a case of feline suicide."

I am not certain how one would go about authenticating such an account, but my friend Marie has told me the story of her older sister, Claire, and Claire's cat. In 1972, when Claire was twenty-two, the family was living in Burgundy, France, southeast of Paris. Claire decided to go mountain climbing by herself in Argenteuil-la-Bassée, an abandoned village in the mountains some three hundred miles away. Nobody heard from her for weeks. The family was worried, and so was her cat, Minou—so much so that one day the cat stopped eating and would not touch food for the next three weeks. A mountaineer noticed a car, Claire's car, that had been parked at the foot of a mountain for three weeks. A search party found her body shortly thereafter. Marie, who told me this story, was at home when the car was brought back. She saw Minou crawl under the car and refuse to move. Nothing could coax the cat out from under the car, and a few days later she was dead. She had stopped eating on exactly the day, it was later learned, that Claire had died. How is this conceivable? Can there possibly be this kind of connection between a loved animal and a human? Is there any mechanism that could explain it? Rupert Sheldrake, who wrote the book *Dogs That Know When Their Owners Are Coming Home*, would presumably say yes. I remain unconvinced but not completely closed to the idea. I can certainly think of no better explanation.

Years ago, I had two Abyssinian cats who were brothers, Rama and Ravana. Rama was killed by a car on one of his solitary adventures. We buried him in our garden in Berkeley. The next day Ravana came and sat on the grave (he had watched us bury his brother). For weeks he would sit on that spot, looking melancholy. He never, I felt, quite recovered from his grief.

My friend Peter Thompson, a filmmaker and writer from Sydney, tells me that his father, the poet John Thompson, lost a cat in circumstances that confirm that cats can die of grief. His beloved Red Ned was put in the care of a veterinarian for several weeks while John was teaching in Malaysia. Two weeks before he returned, Red Ned suddenly died, in spite of being in perfect health and well cared for. The embarrassed vet insisted upon performing an autopsy, but he could find nothing wrong and concluded reluctantly that Red Ned had died of a broken heart.

Surely a cat who is willing to die of grief for you, or to risk her life in following you into what the cat knows is a dangerous situation, will do anything you ask. False. Humans are great lovers of consistency, and it annoys us that cats have different rules. Even if a cat is willing to give up her life for you, that does not mean that she is willing to change her mind at your request. It is a great mystery to me why it is so difficult to get a cat to change her mind. Minna adores me; I have ample proof, even if I doubt she would kill herself if I were to disappear. But when she is about to jump off the bed, and I call, begging her to stay, "Please, Minna, stay with me, Minna Girl, come, come," she looks at me for a moment; a very slight hesitation appears on her face, but she shakes it away, and jumps off the bed. Always. There is rarely an exception: once a cat makes up her mind to do something, your pleadings matter not at all. A dog, in the same circumstances, no matter how urgently he needs to do something, will change his mind if you insist. He is made for compromise, for self-sacrifice, for thinking about how *you* feel. Not cats. That is why I think Minna feels torn for a moment: her whole nature tells her, "Jump, you fool, pay no attention to him. Who is he to you?" However, her experience with me tells her differently. I lose, almost always, but it is a tribute to the depth of her feelings for me that she turns around, looks back at me wistfully, almost as if to say, "I wish I could, but I'm a cat. Ciao."

It is not, I believe, that there is anything else more urgent for Minna, just that it does not occur to her to place my wishes first in her mind. We do the same: we acknowledge love only in forms immediately comprehensible to us. Cats have purring, and a look of love in the eye, and the rubbing against us and the entwining of their bodies around our legs, and even the meow of love, a sound we recognize at once. Why not expand our definition of love? Might there not be forms of love, from cats to us, that we do not recognize because they are not familiar?

In comparing the emotional life of cats with that of dogs, it has been noted (first I think by Darwin's friend, George Romanes) that a dog's emotional life has been modified by contact with humans, whereas this does not seem to be the case with cats. Can we say that cats have not developed emotions simply to communicate with humans? Do they show affection to us differently from the way they show it to other cats? It would seem they do, and they expect us to be able to read their affection. The question is, though, how old is this ability in the domestic cat to "speak" to us so that we understand? Even though the ancient Egyptians considered cats godlike, there is no concrete evidence of affectionate interactions between people and cats. Cats were admired for aesthetic reasons but rarely, it seems, fondled, petted, and made much of. Purring is not referred to in ancient texts. (There is no ancient Greek or Latin word for purr, which suggests that it went unobserved by even the great Greek naturalists—very odd.) This is important, because people may not even have noticed cat purring—that is, they would not have treated the cat as an object of affection, an animal with whom one could be on intimate terms. Juliet Clutton-Brock, the British specialist on the domestication of animals, points out that early on cats would have sat around a fire as a companion. If so, surely they purred? It would be a bold thesis indeed to claim that cats, while physiologically capable of purring, did so only when they became intimate with humans, so late in their developmental history.

If we suppose that early on in their domesticated history, people did not love cats, could cats have nonetheless loved people deficient in this capacity? Alternatively, would they only have discovered love in relation to a human once humans loved them? No doubt cats purred and felt affection for other cats, but it seems unlikely that they would indulge in unrequited love. After all, they are not dogs.

A chapter on love and happiness cannot completely avoid the topic of unhappiness. Alas, I find myself in a position I did not think I would ever come to: I am getting rid of Yossie. I have been with him and I have loved him now for a year. It is shocking that I can think of doing this; I would never consider ridding myself of a difficult child. However, Yossie is clearly not happy any longer. Evidence: A woman with three cats at the top of our hill called to say that Yossie, unprovoked, seeks out her three cats and fights with them. She knows it was Yossie, because he left his collar with his name there. He is also terrorizing another neighbor's two cats. Corina, their owner (she uses the term, I do not), told me that no two cats were ever more spoiled than hers, Lollipop and Captain Puss. Their mother nursed them for six months. (The onset of the weaning period is normally four weeks and is largely completed by seven weeks after birth.) They were born in her house and have never left it for another house, though they are outdoor cats, have never been in a fight, and, Corina says, have never had a stressful moment in their lives. They are contented cats—or were, until Yossie came into their lives. He now squeezes between the floorboards underneath their kitchen and emerges, the cat from hell, to torment them. He bites them, seizes their food, and then occupies their bed. Moreover, as I noted earlier, he scratched Ilan unprovoked. Yesterday he lashed out at a three-year-old girl who was playing at our house. He also has begun to bully the four smaller cats. When they try to play with him, he plays so roughly that they scream for help, especially poor Moko, who for unknown reasons finds himself branded as enemy number one. Yossie is making his life hell. He will not come on walks with us anymore, just stays home and sulks. I called Jane, who gave him to us originally, and she has agreed to take him back to her house with her 120 cats! We will see how he does there after ten days and reevaluate the situation.

What would account for this gradual change for the worse in Yossie's personality? I think an earlier pattern is emerging. Yossie was a stray, and there is no such thing as a stray cat who has not led a difficult life. It may well be, thinks Jane, that children tormented Yossie when he was a kitten, and it is all coming back to him now. Why it took this long, and why he was so content for nearly a year, is a puzzle. He did not seem to mind at all when the other cats joined the household. He was ever tolerant. Since we moved to the beach, however, he has not been the same. Maybe it was just one move too many. Maybe he really is a cat who wants to live an unfettered life. He is just not a joiner. How will he fare with 120 other cats? We will see, but I have a feeling he will seek out his own territory, perhaps the one he held before, and will feel less pressure to be a household cat. The point is that at the moment he is not happy. Or rather, this could be a moot point: perhaps he *likes* doing what he is doing, but it is making other people and other cats unhappy, and we cannot give priority to his own wishes over everyone else's.

Alas, he does not possess what the nineteenth-century German psychiatrists called *krankheitseinsicht*, insight into the fact that he is ill. He does not know that his time is limited here. He does not brood and think to himself: "I have brought this on myself; I have nobody to blame but myself." Unfortunately, he will probably never know, nor understand, the drastic change that will take place in his life. Like the parent of a problem child, I cannot help wondering if I did not do something wrong to create this situation, if I could have avoided it by doing something differently.

News bulletin: Ilan, on a walk this morning, asked if I would not reconsider. Clearly he identifies, with good reason, too, with Yossie and his impossible behavior. "Papa," he told me, "please give him another chance. Look how good he has been today." It was true, too. Yossie had been grooming each of the cats, one by one, and had accompanied Ilan up the hill by himself yesterday evening. I could swear he knows what is up and is trying to avert the disaster. Well, a reprieve is not out of the question, and I agreed with Ilan that he would get another week, and then we would see if he really was beginning to change for the better. (The week has now stretched to two months, and the verdict is still out.)

It is difficult for me to admit, but it may well be that some cats simply do not need other cats, or other people, to feel happiness. Cats have been solitary for so long in their evolutionary history that other animals, especially humans, may be irrelevant to their emotional lives. Their happiness was found, forcibly, outside of a relationship with another cat. Humans think of the happiness that is at issue here as being superficial, purely material. It is almost impossible for us to accept because it is foreign to our own experience. I know that if Miki simply moved next door (he likes three or four of our neighbors and spends considerable time at their homes), I would be shocked and hurt, but not entirely surprised.

Do cats fall in love? All mammals, of course, are "attracted" by many of the same things we are: a fit body, good skin (sleek fur), shining eyes, everything that indicates health. However, this attraction has a purpose: to find a healthy mate for reproduction, or so evolutionary biologists will tell us. It appears true for cats. The relations between male and female cats strikes us as purely sexual, having more to do with lust or, more accurately, reproduction than love. This is because the attraction does not seem to extend beyond the few moments needed for mating. In none of the cat species, with the exception of lions (and perhaps cheetahs), does the male cat play any role in raising kittens. Nor do male and female cats who have kittens in common seem to recognize any special bond. Humans have a deep need to define love by some evidence of permanence. This could be our own particular prejudice, or sexuality may simply play a minor role in the love life of cats.

What about the love of a mother cat for her offspring? Sue Hubbell, who writes so convincingly about the life of bees, tells of a beautiful but neurotic and twitchy calico cat she lived with. The cat was neurotic, she explains, because the cat's mother, who lived with Sue Hubbell as well, "batted and hissed at her daughter every time she saw her, which was often." We can presume that this was after the daughter was an adult cat. We naturally expect a mother cat to continue her love for her offspring for her entire life. Since this does not happen without fail in humans, why should we expect more of every cat? Many mother cats adore their daughters and choose, even in feral circumstances, to live with them permanently. If mother cats do not show the same love for their sons, and indeed often chase them away after a certain age, this could well be because male cats seem to have no concept of incest. Personal likes and dislikes, however, seem to play as strong a role in the lives of cats as they do in the lives of humans.

Who do cats think we are? Many people have said that cats regard us as another cat, some say a supercat, others a defective one. I think this is wrong. Cats know perfectly well that we are not cats; we do not look like cats, and we do not behave like cats. Nor do we smell like cats! (Notice how cats love to smell each other, especially around the anus. They do not do that to us.) They do not think of us as gods (witness how often they ignore our presence and demands). They do not see us as enemies (except when we are). They do not recognize the category of master or owner. (There is a recent campaign, started by the Mill Valley group In Defense of Animals, founded by the veterinarian Elliott Katz, called "We are not their owners, they are not our property," asking that people no longer see themselves as "owning" another living being. I am in complete agreement with their goals, for the idea that one person can "own" another being, human or animal, belongs to the history of discarded ideas.) The idea, to a cat, that somebody else owns him is ludicrous.

There are people who believe that cats regard us as their mother, and I have seen some indication of this. Now, however, I am beginning to wonder whether we may not have it back to front. Maybe cats see us as kittens. Some evidence (possibly too large a word for this) suggest there is some truth to this notion. When a mother cat approaches her kittens, she usually does so with a "brrp" call. Whenever my cats see me after an absence, this is the sound they make. (It is also called, variously, a mrhn, murmur, or a chirp but is different from the ordinary meow.) A mother cat is vigilant over her kittens, and when any kind of danger, perceived or even imagined, is near, she will give a sudden growl to warn them. I often wake at night to see one of the cats standing (protectively?) over me, looking out at the deck next to our bedroom, and making that same growl. I have always assumed the cat was addressing the unknown assailant, but now it occurs to me that the cat may be addressing me, warning me to be careful. The cats are definitely not talking to themselves; the sound is meant to communicate. Of course, many people have wondered whether the unwanted offerings of dead mice and birds that cats bring us so often is not identical to the prey that a mother cat brings to her kittens, both to feed them and to teach them to hunt. In other words, they are trying to coax us into being better cats. The problem with this neat hypothesis is that male cats, who also bring offerings to us, do not act in this protective way toward kittens—although they are tolerant of kittens in ways they never are of grown cats. (My own theory about why cats bring food home is that it is a good place to keep it safe from competitors.)

Perhaps there is a difference in how female and male cats treat humans, a subject that has not been investigated and might yield interesting results. Toms (before they are neutered, in any event) tend in my observations to see men as competitors. We often speak of female cats as flirtatious, but I have yet to see a difference in the way they treat men or women. If our cats seem more attached to me than to Leila, it is surely only because I pay more attention to them than she does. A female cat rolls on her back when soliciting sex. When Minna first did this, I was alarmed. Now, however, I see the males do it to me as well. It must be that kittens do it to make their mothers give them what they want.

We expect cats to be selfish and egotistical. Does love make them otherwise? Can a cat exhibit compassion, for example? I have never found a single account of compassion in cats in the scientific literature, although I have been surprised at how many reliable stories of the compassion of cats I have heard.

I have a close friend whose cat, known as Fu, *the* Fu, had only three legs. Like most animals, Fu seemed unaware of her disability. I knew this cat well; she struck me as a cat with exceptional poise and even—dare one say it for a cat?—charisma. My friend told me, though, that she was also capable of displaying remarkable empathy. One night when my friend was very upset, Fu did something she had never done before—she crawled up on the bed and laid herself down on my friend's chest and purred and stayed there all night. Months later, my friend was about to undergo surgery to have a breast lump removed. She was anxious and frightened. Fu again did something she had never done before—with her three legs she accompanied my friend all the way down three flights of stairs to the front door of their apartment building and then stood there talking and talking to her until my preoccupied friend noticed she was there.

Victoria Thompson, a writer from Sydney, told me that her cat, Mish Mish (Arabic for "Apricot"), would come to comfort her when she was upset. "She would walk into the room and if she found me weeping, she would stop in the doorway, looking at me intensely with her deep blue eyes. A look of concern would come over her face as she approached me. Her little paw would dart forward and touch me as she stared into my eyes. She would blink once or twice with affection, before settling quietly beside me to keep me company, perhaps to watch over me, too."

Many years ago I had a cat called Megals, who was clever and devious in ways that one does not usually associate with cats. Our family had just acquired a young poodle puppy, and we were walking outside, the cat in front, when suddenly Megals began to run very fast, turning back to look at the puppy (Misha was his name), inviting him for a chase. Misha complied and began running after Megals, when suddenly Megals dug in her claws and stopped abruptly. It was too late for Misha, who proceeded to fall down a steep embankment. I suspect, though I was not certain, that Megals knew what was coming: it was his idea of a good joke. We all ran to the edge of the small cliff, including Megals, to see Misha lying in a heap at the bottom, crying more in bewilderment than in pain.

I was about to rebuke Megals for what I regarded as a dirty trick when he suddenly leapt down the cliff and landed next to Misha, whom he proceeded to lick and rub against. It looked to me like an apology, driven by compassion, but I may have been fooled.

Altruism, the close cousin of compassion, is not entirely absent in cats, but most accounts relate to mother cats, who are famously attentive to their kittens and will often risk their own lives to save their children. Many in New York remember the story of Scarlett, a calico cat who was in an abandoned garage in East New York that was set on fire on March 29, 1996. David Giannelli, one of the first firefighters to arrive on the scene, heard a strange sound and found three small kittens on the grass, in a line. Then two more. Then a few feet away he found the young mother. "Her eyes were blistered shut, her mouth, ears, and face scorched, her coat badly singed. Her paws were crusted with thick black soot." However, she was alive. This street cat had raced back into the blazing garage, running over burning embers five different times, to save each of her five kittens. When David reunited her with her five kittens he had put in a box, "she made a sightless head count" and then began to purr with relief. He drove her to the North Shore Animal League in Port Washington on Long Island. Her inner lids were badly swollen; her skin was severely burned; some of the hair on her body had been singed off; the pads of her feet were burned, as were her mammary glands and her ears. Would she live, and if so, would she ever see again? Even in that condition, when she was put with her kittens in a special oxygen chamber, the first thing she did was count them with her nose, then lick each one, and finally rolled onto her side to let them nurse. Her courage, her pluck, and her willingness to allow herself to be cared for, as if she were saying, "I did my work, now you do yours," amazed Dr. Larry Cohen, the veterinarian who took care of her. He did do his best; she survived, as did four of the five kittens, and the world learned about the depths of nonhuman maternal devotion.

Can a cat fake love? The idea seems ludicrous. Hypocrisy is one character trait that you will rarely find any cat accused of (although Colette famously speaks of her cat displaying a "particular kind of honorable deceit"). We find it hard to believe that a cat could pretend to like us, without feeling much of anything for us—that is, that a cat could feign affection to achieve some other end. Manipulativeness, on the other hand, is fairly common in cats, as when my five cats wrap themselves around my legs at feeding time. Yes, they are showing me affection, but yes, it has a specific purpose: to get me to feed them even faster and to make certain I do not change my mind. It is also an indication of their gratitude, to some extent, and anticipatory pleasure. It is not hypocritical, but it is brazenly manipulative.

While food is appreciated, it is clearly not the only thing that brings out this behavior. Usually it means that they want something; for example, they may rub against me to convince me to open the door when they want to be let outside. That is only similar to a young child saying, "Please please please," to get what he or she wants. Were this the only time the cats did this to me, I could begin to doubt their sincerity and wonder whether they really mean what they are showing. But then Minna, in particular, will do the exact same dance at my feet from the sheer pleasure of seeing me and even will do so several times within one hour. The meaning is entirely clear: "I am so glad to see you."

In fact, I sometimes wish that some of the cats would be a bit more hypocritical and show me affection even when they do not mean it. They will not. For reasons I understand not at all, Miki has chosen, quite deliberately, to be cool to me lately. He likes to sleep at the foot of my bed, but he sleeps at the head of Ilan's bed. If I pick him up, he jumps out of my arms immediately. If Ilan picks him up, he stays. He rarely acknowledges my calling him, but when Ilan calls, he usually comes running. I think it is clear: He prefers my son to me. On the other hand, he will never pass up the chance to go for a walk along the beach or up the hill with me. If Ilan or somebody else goes, he will rarely follow, but when he sees me head for the door, he becomes alert and rushes to accompany me.

Some of this has to do with how the cats assess the situation. For example, Moko, who dislikes having his head patted, knows that he cannot be as rough with small children as he is with me. When I do it, he swats me in a kind of rebuke. He knows I know that he dislikes it, so why do I insist? If a small child does the same thing, he tolerates it. He never scratches a child, and he does not run away. It is clear to me that he does not enjoy it, but he seems to have made up his mind to grin and bear it, because of the age of the child. He *knows* it is a little child, and either he does not want to hurt the child's feelings—admittedly a sophisticated notion that is not easy to justify, but is worth thinking about—or he simply makes allowances for a child that he would not make for me. I am not hurt; he has a point, after all. Megala does the same; he allows himself to be handled, even manhandled a bit, by young children, but never by me.

Oddly, Megala too has become less tolerant of my affection. As he grows older, it is as if he finds my demonstrativeness embarrassing. However, here is the strange thing: It is geographically driven. In bed, he will not allow me to pat him for more than a few seconds. But when I sit, as now, at my computer, he will not stay off my lap, and at that time, and *only* at that time, I am permitted to do anything at all to him—in fact, I am commanded to do everything to him, stroke him, clean his eyes and ears, play with his claws—while he purrs and pushes his head against my chin, sounding and behaving just like the leopard his ancestors were.

All of this is the opposite of hypocrisy. While most humans would be embarrassed to withdraw affection so easily and in such an idiosyncratic fashion, cats have a take-it-or-leave-it attitude. Their behavior says: "This is what I feel; would you rather I pretend?" If dogs do not lie about love, cats do not lie about affection, even if we expect them to and want them to. The little courtesies we use to make life smoother are foreign to the feline nature. It is just not worth it for them to pretend.

Cats have been, since antiquity, ready and willing to meet us halfway to love—they have always been prepared to show us all the signs of love and were just waiting for us to ask. It has taken humans a long time, but it seems we are at last responding to the signals cats have been emitting, like some lonely inhabited planet millions of light-years away, lost in space and hoping for companion stars somewhere in the universe.

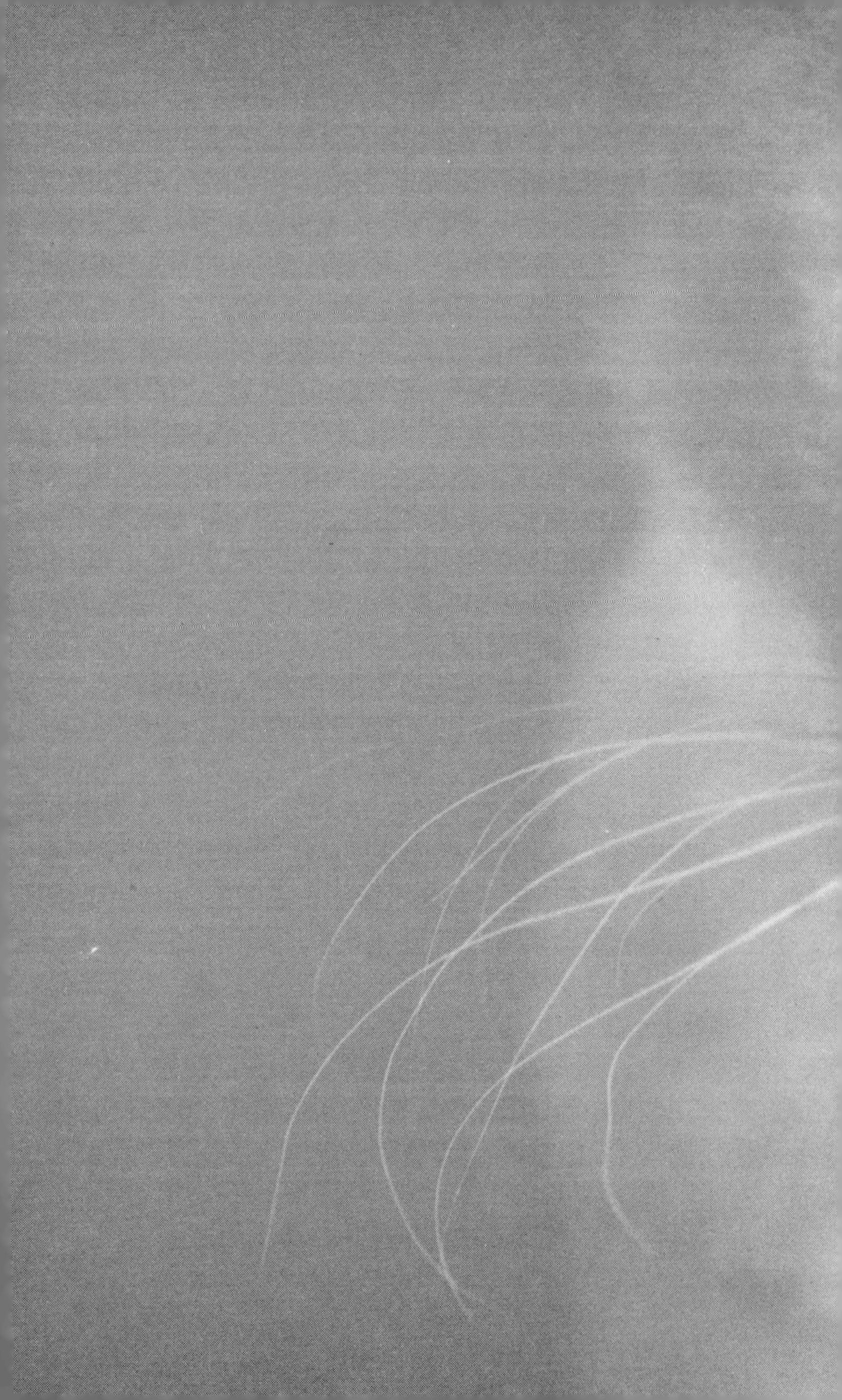

THREE

Contentment

Moko

I am always amazed at what skittish creatures cats are. That is surely why they usually do not like to be around small children. Miki, but only Miki, does not mind Ilan picking him up, banging drums in his presence, ambushing him, jumping on him, and all the other things that impossible five-year-old boys will do. The other cats are less tolerant. Moreover, in the beginning when Ilan's friends came over and did the same things, all but Miki would leave the house immediately (now they are more relaxed, since they know there is little to fear—somehow they realize the children mean them no real harm). They do not like unpleasant surprises. Why should they? For jungle creatures, who must always be attentive to danger, especially from an unexpected source, any sudden movement can spell deep trouble. At night, their senses are particularly vigilant. They hear and listen to every noise. They are on the alert. They are, without question, anxious creatures. They evolved to be so, and it is in their self-interest to maintain alertness at all times.

I think it is fair to say that cats love security and hate to be vulnerable. When they feel secure, and not vulnerable, we can say they are content. This contentment is a particular, even a peculiar, emotion—distinct from happiness—and I think it is one that cats, of all animals, demonstrate most clearly. One might argue that it is nothing more than the absence of anxiety. However, many animals appear to be without anxiety and yet we would not consider them content. The cat does not merely experience contentment, he exudes it. You cannot be in the presence of a contented cat and not have some of that contentment rub off on you. Which surely is a good part of the reason we love cats so. Scientists have observed that petting a purring cat lowers our blood pressure; I believe that it can raise our morale at the same time. Why is this? To feel needed and appreciated by a creature who seems so self-possessed, apparently needing nothing beyond his own perfect self, can be very cheering. "I can't be all bad, if my cat likes me so much."

Sometimes, I confess, I am not certain if my cats are content or experiencing a kind of boredom or listless inertia. So it would appear from the outside. Yesterday, after a day of much activity, long walks on the beach and a walk in the rain forest behind the house, the cats sat about in the evening looking blank. I mean, of course, that is how it looked to me, perhaps because all five were spread out in different parts of the living room, not seeking one another or our company. Miki was sitting like an Egyptian statue of a cat, his eyes half-closed, listing slightly to one side. I put on Bach, for sometimes they seem to enjoy music. (David Morrell, a novelist from Iowa City, claims that his cats love opera and "all gather as a clan and sleep in front of the speakers for eight hours. They especially like Puccini.") No response. Normally they perk up, wander closer to the source of the music, and then settle down on the sofa. Were they content, were they bored, or was their apparent torpor merely a sign of exhaustion? When it comes to these more subtle emotions (as opposed, say, to anger or aggression), we have problems finding our way into the minds of other creatures.

Purring is almost always a sign of contentment (although there are exceptions noted below). Unlike a human smile, purring cannot be, as far as anyone knows, faked. I have never heard anybody say about a purring cat, "She is only pretending." Neils Pedersen of the University of California at Davis School of Veterinary Medicine explains that experiments have shown purring to originate from within the central nervous system and is a voluntary act, not a reflex.

Nobody is certain of the mechanism involved in purring. A cat can purr and make other sounds at the same time, since the purr can be produced with the mouth closed. Nursing kittens begin purring when they are a week old, an indication to the mother that all is well. Because purring is a vibration rather than a sound, it protects the kitten from predators yet lets the mother know that the kitten is not in any danger. She responds with a purr of her own, a response in a positive feedback loop. Do cats ever purr without a living being present? I do not think so, though surprisingly I have not been able to find any studies done with hidden cameras and sound equipment that investigate this. If I am right, then purring is a form of communication, and what is communicated is contentment. (I have suggested that dogs, too, wag their tails in only the presence of another living being, because wagging a tail is also a form of communication. Similarly, humans cannot tickle themselves, because tickling involves an interaction with another person.) Cats do not purr *by* themselves, which would seem to mean that they do not purr *for* themselves, either, but for us and for each other and even for other animals they like. I have not heard of a cat purring in the presence of a favorite toy. It may well be that every cat has his or her own distinct "signature purr," recognized by other cats: "Oh, it's you." Some cats also purr when they observe another cat's pleasure. Once when I was cleaning Yossie's ears, Minna began to purr. Either she was imagining the pleasure he felt or his pleasure gave her pleasure—or maybe it is that purring is, like a yawn, contagious. Purring is so essential to cats that the very earliest reference to purring in the English language makes it sound as if purring were the very essence of cathood:

The cat amid the ashes purr'd,
For purrs to cats belong.

Several of my cats sleep under the covers with me; when they first come to bed, they are purring madly, I assume at the pleasure of being with me and with one another. As they fall asleep, the purring diminishes, and by the time they are asleep, the purring has stopped. I have observed them dreaming (you can see the REM under their eyelids, and their paws move as if they were running), but I have not noticed that they purr when they dream, which is surprising. Would they not sometimes be dreaming of pleasant experiences, and if so, should there not be times when dreaming cats purr?

Cat daydreams, after all, appear to trigger purring. Minna will sit on the edge of our hot tub in total silence. Suddenly, for no discernible reason, she will begin to purr. It must be that she has imagined some pleasurable situation, and her response is to purr. I am present, so the purr could in theory be directed at me, but since her eyes are closed, and she is not paying any direct attention to me, I believe she is in the throes of imagining something wonderful. The question remains if she would have begun to purr even with no audience. I suspect not, but I am still puzzled by the absence of purring in dreaming, where an audience is at least imagined. Maybe cats have lucid dreams—that is, dreams in which the dreamer knows he is dreaming. They sleep remarkably lightly. They can hear the faintest sound of a mouse's footsteps and be instantly awake and alert.

Do cats purr when they are fed? Only, I have discovered, when we are present. Minna was thrilled when I fed her a favorite treat the other day and began to purr wildly as she ate it—until, that is, I left the room. I could hear, as I left, the purring cease. I do not know how to interpret this except to see the purring as talk, her way of thanking me.

Cats have other sounds they make to express pleasure. Kittens, including wild cat cubs, use vowel sounds, something that adult wild cats never do. These vowels—*e*, *o*, and *u*—generally express pleasure. Therefore, when adult domestic cats meow, using vowels, they are behaving more like kittens, again a result of domestication—neoteny, in a word. (Neoteny is when an animal retains juvenile characteristics into adulthood, both at a physical level—round face, large eyes—and at a social level—friendly, approachable, gentle.) In the wild, too much vocalization would call attention to themselves, alerting a predator to their presence. Wild cats cannot afford to be vocal; our cats can. An adult wild cat does not need expressions to convey pleasure to another cat; they come into contact with other cats so rarely that communication is pretty much restricted to negatives, like "I don't like the look or smell or sound of you," which are easily conveyed by growling, hissing, and other bodily gestures. Wild cats, then, would appear (it is not known for certain) to purr only in the rare instances when a male and female are together for purposes of procreation and, of far greater importance, when the mother cat has kittens.

Desmond Morris, one of the world's leading cat experts, maintains that all small cats (including ocelots, servals, and other wild cats) can purr, but the big cats, the lions, leopards, and tigers, cannot. They roar, but they cannot purr. Not everyone agrees, however; some insist that all cats purr. Elizabeth Marshall Thomas, the best-selling author of books about both dogs and cats and a discerning anthropologist, has heard a leopard purr. She also has noticed that her cat purrs extra loud when he is with her husband. Her husband is hard of hearing, and it appears the cat recognizes this. If this is true, it confirms my opinion that purring is a form of communication, similar to a human smile.

Veterinarians know that cats also purr when they are in distress. The predominant explanation is that they are self-medicating (the purr of a cat is a powerful tranquilizer, for cats and for humans). We can liken it to when we hum or sing a tune to ourselves, which we also use in moments of distress. Animal behaviorists also note that cats purr when confronted with strange cats; it is surmised that they do so as a form of submission, since cats know that a purring cat will not attack. A distress purr may also be a solicitation signal, a way of saying, "Please help me." A mother cat giving birth will often purr. Is she anticipating the pleasure or masking the pain? Vets have told me that when they have to euthanize a cat, the cat will often begin to purr. Is it just that the vet or assistant is holding the cat? Or is the cat using purring as a form of begging? "Please don't kill me; look how much pleasure I could still bring you." It breaks one's heart.

Purring is the external manifestation of an internal feeling. It is a visible sign of contentment. Another is the nictitating membrane, or third eyelid, something the cat has that we do not. This third eyelid is normally folded into the inner corner of the eye (this membrane is also found in sharks, owls, and polar bears, who use it to prevent snow blindness). Like a white blind, it sometimes moves over and covers the cat's pupil when the cat is dozing. When first seen, it is somewhat frightening, as people can think their cat has suddenly developed glaucoma. It is wonderful protection for the eye against getting scratched when cats are stalking through tall grass. When a cat is very comfortable and being petted, this third eyelid moves over the eye. I am not sure of the significance of this; perhaps the cat is so relaxed that he has let down his normal guard, and the physiological response of moving into defensive posture is triggered automatically, the body's way of saying to be careful.

What, though, are the conditions that bring about this positive state? It cannot be only the absence of anxiety that makes a cat content. There is also the matter of trust. You cannot be content if you do not have trust in your immediate environment. When the cats stretch out on our bed, rolling over and reaching out their paws to us, this luxuriating trust conjures up in us a sense of accomplishment, that we have behaved in such a way that another creature trusts us completely. I am not arguing that this confidence is well placed. After all, a cat could feel this trust in someone who the next day casually drops him off at the local shelter with the request that if he cannot be placed somewhere, he should be "put down."

Moko is resistant to being carried. Actually, most cats do not like it. Being carried involves too much dependence on another person; never a cat's preferred MO. Yet every once in a while when Moko meets me on the path down to our house, he reaches up to me with his paws, scratching my legs as if asking to be carried. I always feel this is a special privilege he has accorded me. I am careful to support his whole body when I hold him so that he does not feel unsafe. I also try not to impose longer than he wishes; as soon as he makes it known, by trying to squirm out of my arms, that he has had enough, I gently and quickly put him down on the ground. This is the moment when anxiety is absent for Moko; he is content in my arms. I am building up a fund of trust. He is finding me reliable, and I feel honored.

• • •

Are cats ever more than happy? Are they ever manic? If you have a certain plant in your garden, *Nepeta cataria*, popularly known as catnip, you are likely to think so. This Asiatic plant is related to the marijuana plant and contains the chemical nepetalactone in its stems and leaves. Cats rarely attempt to eat the plant but will instead sniff it, or rub against it, or roll on top of it. They will chew it, bite it, and lick it as well. As explained by Desmond Morris in his feline encyclopedia, *Cat World*, they may go into a trance or into ecstasy or become contemplative. Some cats are enthralled; others go wild with pleasure, purring loudly, rolling over and over and even leaping in the air. The disinhibition affects all cats who respond, but not all cats do. Strangely, none of my five cats react to the catnip in their toys, and they react only mildly and briefly to the plant itself. In fact, about half of all cats have no reaction at all to catnip. The propensity seems to be passed on genetically. Evidently catnip oil mimics a feline reproductive pheromone, which is probably why kittens are not affected and why all kittens avoid catnip; it is only at about three months that half of them show the intense reaction. Males and females are affected equally, so it is unlikely to be related to sexuality (it is not an aphrodisiac, as popularly thought), though the cat's response looks like that of a female in estrus. Neutered cats respond as well. Catnip appears to induce a state in cats similar to the one humans induce by means of marijuana, LSD, and other mind-altering drugs. Is this happiness? That is a complicated philosophical question. Of course, the reaction is intense, but perhaps out of prejudice, I like to reserve the word *happiness* for moods produced more naturally, such as feeling happy that the sun is shining or that you are playing with your friends on the beach.

In English, at least, if not in "cat," the word *contentment* conveys something of a feeling of being at peace with the world or with yourself. It is more of a state than a fleeting emotion. A person can be happy (momentarily) without being content. Contentment cannot be purchased; happiness, on the other hand, has a price. For us, happiness is serious business. Our own, other people's, and our companion animals'. The proof is in the amount of money we spend per year on pet products: more than $7.5 billion on food and almost $4 billion on accessories in the United States alone. We consider the question of what is happiness for a cat not to be a trivial one, either at a practical or a philosophical level. It is a serious question: What gives a cat pleasure? What makes a cat happy?

Happiness is an active emotion, something that comes and goes and which we humans seem to control, or believe we control. Contentment is not subject to our will, not something we have all that much influence over, and in the case of cats, it seems to be almost a totally separate emotional condition. A cat's very posture can convey contentment. Everyone has seen a cat sitting in a compact manner, her legs and tail tucked neatly underneath, eyes half-closed, and sometimes even purring (if we are present!) in ecstasy, or at least contentment.

When we use this expression for humans, we tend to do so in a somewhat negative manner, for it suggests smugness. Here is where we might take a page from the emotional lives of cats to our own benefit: their ability to feel the happiness or contentment of the moment is to their glory, just as the ability not to feel happiness when our neighbor is suffering is a human trait we encourage. It is hard to feel intense joy when someone we know is in intense misery. Cats, of course, never worry about this, *as far as we know*. I lay emphasis on this qualification, because while it seems a fair assumption, it is one that we make with very little knowledge. We just do not know how the suffering of their friends affects cats, if it does. We assume that it bothers them not at all, and the evidence is constantly before us. Nonetheless there are instances, anecdotes, of cats who are clearly worried when their friends are unhappy, some of which I've repeated here. I have been flooded with stories of cats who know when their human companions are ill, even down to the painful organ. They will sit for hours on a sore leg or wounded arm, purring away as if they expected the vibrations to perform the miracle cure that it sometimes seems to achieve.

It could be argued that humans are the only creatures who can become unhappy at *imagining* the sorrow of another creature, as opposed to witnessing it. I seriously doubt that my cats would become depressed just thinking about what might happen to me when I am away from them. In spite of my serious doubt, let me also express my ignorance: Some cats appear to become sad when we are absent. We presume, and it can only be a presumption, that they are suffering because of what *they* are deprived of, not what they imagine is happening or could happen to us. In other words, their concern is self-centered. This may well be true, but we cannot know for certain that their imagination is not running away with them. I may be projecting when I think my cats are worried when I go swimming, but it could also be true. They are more than curious, as I pointed out earlier; they are interested and attentive. Should I suffer a serious accident in the water, I know that none of them will attempt to come to my rescue. Unlike a dog, of course, they cannot help me. They could hardly drag me out of the water. Saving a drowning human being is simply not part of their genetic heritage or their experience, and I suspect not even part of their imagination. Still, I do wonder what they are thinking about when they cluster together, as they do, on the shore, waiting for me to emerge from the water. Today when I came out, Moko gave a little prance of joy, kicking his back feet up in the air, and I interpreted this dance to mean, "Yay, you're safe."

When I first began swimming, once the weather and water were warm enough, the cats accompanied me to the water's edge. Curiosity, I thought. Now my daily swim is like the walks we take. All I have to do is say "swim" and the cats come running to the door. They accompany me every time to the ocean, and if I decide to go back in again a few minutes later, they come back with me. All four of them clearly love sitting at the water's edge, daintily reaching out their paws to touch the waves as they lap on the shore and watching me swim out to sea. As I begin my swim back, they become more boisterous, and they leap over one another and wrestle in the sand. Miki will grab a seashell in his mouth and toss it in the air, then grab it again and carry it up and down the beach. This is something that is surely related to contentment but is different, more like joie de vivre. All, except Yossie, have it, express it, and live it. (Whatever Yossie suffered in his early life as a stray seems to interfere with his ability to experience joy.) It is different from the joy they get from being able to do what cats do well (hunting, climbing, jumping), more like an externalization of their contentment, a sign to me perhaps of how good they feel their life is going for them at this moment. It makes me very proud.

Elizabeth Marshall Thomas, in one of the best passages of her book *The Hidden Life of Dogs*, says that what dogs want is to lie quietly in the shade on a lazy sunny afternoon. "While the shadows grew long we lay calmly, feeling the moment, the calmness, the warm light of the red sun—each of us happy enough with the others, unworried, each of us quiet and serene. . . . Primates feel pure, flat immobility as boredom, but dogs feel it as peace." A cat lying in the sun—as in the marvelous photos by the German photographer Hans Silvester in the book with that name (*Cats in the Sun*), about felines and sunshine in the Greek islands—is certainly the picture of contentment. Are we part of this pleasure? If we were lying next to the cat, would the pleasure be any greater for the cat? I like to think so, but if cats evolved as solitary hunters, what do we bring to them that they do not have on their own? Well, there is an answer to this question, and that is that while the *ancestor* of the cat evolved to be a solitary hunter, our domestic cats have not. They evolved to live in a kind of symbiotic relationship with humans. We provide them food, shelter, and affection, in return for their companionship and anything else they might offer us (rodent control, winter blanket, stress reliever). Many people feel more complete with a cat in their life, and I would not be surprised if cats felt the same way about us. I know that if I disappeared from the lives of my five cats, they would not be as happy as before. I know, because they wait for me to go on walks along the beach, though they could perfectly well go on their own. When I am with them, they react in such a strong way, gamboling, racing ahead of me, and then flopping down in my path, that it is obvious they derive great pleasure from my company. I find it hard to believe, though, that they could possibly enjoy my company as much as I enjoy theirs. This is not surprising: we domesticated cats for our benefit. While they get something from it, we probably got the better deal.

It wounds our vanity to think that cats can be content without us, so much so that when we look for positive characteristics of a cat's personality, high on that list is friendliness to humans. That may be very important to us, but it's far less important to cats. Their contentment may be independent of this trait, much as we like to think of it as essential to a cat's happiness. If we insisted on this point, however, it would mean that there could be no such thing as a contented solitary cat, or indeed any cat living by himself in the wild. This would be a foolish statement, though clearly very little thought has been given to the sources of happiness in animals who live content without human companionship. Indeed, a common complaint from people who do not like cats is that cats are perfectly content *without* humans. This might well be true, but it does not rule out that we add something to their contentment.

What allows cats this mysterious contentment, this emotion that resembles ours but seems somehow deeper, more permanent, and even more authentic? Cats seem not to think about the future, even the immediate future. It is as if feeling the full emotional force of the moment absorbs all their psychic energy. They are alert to danger, but they do not anticipate disaster, even death, with worry. Nor do they look backward with remorse or feelings of guilt for what might have been, if only they had been different. They somehow feel themselves as perfect beings. Maybe this is why Zen monasteries often have a cat in attendance. There does seem to be a Zen of cats—a state of self-absorption that is without a trace of self-deification, just an acceptance of some natural order where all is right in the best of all possible universes, a mixture of peacefulness, tranquillity, safety, sun-filled laziness, and the joy of knowing life is good. It is contentment of the most natural kind, so natural that it sometimes appears to exist on a higher plane. This is perhaps what cats are here to teach us: to live in the moment so completely, with such absorption, that the moment lasts forever.

FOUR

Attachment

Moko, left, *and Minnalouche*

"He walked by himself and all places were alike to him." So goes the famous refrain from Rudyard Kipling's story "The Cat That Walked by Himself." Kipling called cats in that story the "wildest of all wild animals." Yet here is the great puzzle: Descended from an entirely wild and elusive ancestor, one mortally afraid of humans (with good reason), the house cat, the domestic cat, our cats, evolved in a mere blink of evolutionary time, to be able to form one of the deepest and most mysterious bonds between human and animal ever recorded.

It is not much of a mystery as to why we love cats, and the answers have come from every person who has ever adored a cat, ever since cats have been companions to humans. We get mousers, companionship, warmth, and mystery in our lives, and a sense of living with a foreign species, even a wild species (no domesticated animal is less domesticated than a cat). Why cats like us is slightly more elusive, and there are many theories. Nobody knows for sure, so we must call them just theories. Heading the list, though, is the simple word *companionship*.

Yes, cats evolved to be lone hunters, but living in proximity to humans for some four thousand years has changed that, irreversibly. I have observed feral cats in cat colonies in America, in Italy, and in New Zealand. Most estimates suggest that there are as many feral cats as there are household cats, so in America alone there are probably fifty million feral cats. The feral cat colonies I have seen are in cities and contain generally a dozen or so cats. What strikes me is how carefully tuned in these cats are to the doings of the humans they encounter. Some people walk by them in indifference; others feed them with compassion. The cats take it all in. You can see their watchfulness, their careful scrutiny of each person who comes near them. As a newcomer, I saw some cats approach me with great hesitation, perhaps terrified that I might hurt them or turn out to be unreliable. In any event, I was unknown. The cats were not indifferent to me, however; quite the contrary. I was either friend or enemy, but it mattered to them which I was. Even feral cats are still tuned in to humans. They are attached to us, even if only from a distance.

One can make too much of the solitary nature of the cat. I am amazed to see how quickly my four young cats have come to depend on one another and to take pleasure in one another's company. Lately they are there waiting for me at the top of the hill when I return from my errands, an hour or sometimes two or three hours later—all four or even five of them. They rush forward when they see me, their tails held high, all making the happy greeting sounds, the birdlike chirps, I love to hear. What do they do while I am gone? Whatever it is, I am sure they do it together. They are not littermates, they were not raised together from birth, but clearly they enjoy being together. They act, almost, like a pack of dogs.

For years I was fond of Abyssinians because they had the reputation of enjoying going for walks with their humans. I had five, in succession. Their reputation was well deserved: all five walked with me. At the time, I was living in Berkeley, California, and I knew few pleasures as great as walking with my dog and cat in the hills behind our house. The problem was that the cat flap also allowed my Abyssinian (Yogi was the first) to continue his walks without me. Invariably this proved fatal to the cat. Eventually, the cats who walked with me came to a bad end, usually hit by a car. That is why almost all vets will tell you not to leave your cat out at night (agreed), and many will tell you not to allow them unsupervised visits outdoors as well. I understand. I just cannot bear to confine a cat. In that sense, cats are unlike dogs. We can provide dogs with all the access to the outside world they need simply by going on enough walks with them. Good for the dogs, good for us. Not so a cat.

It is true, of course, that cats do not need to be walked,

like dogs. So living in a high-rise apartment is easier on a cat than on a dog, who may get one or two walks a day when he would like a dozen. However, it is not ideal. This is why so many people go to extraordinary lengths to enrich their cats' lives, to make their houses interesting to cats. Buying cat toys and cat scratching posts and little houses is only the beginning. You can get carried away (and is not that one of the great abilities we humans seem to possess over and above our companion animals, that we indulge in excess? Well, dogs seem to indulge, too, but cats rarely spoil us to the same degree that we spoil them—or am I being unjust here?) and turn your house or apartment into one that is arranged entirely for the pleasure of your cat. Walkways from room to room; scratching posts that go from floor to ceiling; hiding places above anyone's head. I had the pleasure of seeing firsthand a home like this, described by Bob Walker in his book, *The Cat's House*. However, no matter how much you do at a physical level, the one thing every cat most longs for is still only the small flap that is the cat door.

Cats need access without us. I have titled this chapter "Attachment," but I do not mean just the attachment to human contact. I mean the attachment to the external world altogether. Cats may not have evolved to live with other cats, let alone human beings, but they certainly did not evolve to live their entire lives deprived of an external world.

• • •

On a glorious, warm, sunny winter morning, I decided it was time for the four kittens to explore their beach. They had not yet been outside. I opened the door, and suddenly they could see the ocean. They stopped, looking struck: sun, air, and scents. They sniffed, they stared, and they jumped in the air. They seemed to be letting me know that this was what they need: freedom. The desire for freedom, whether a state of mind or a feeling, is as essential in a cat's life as it is in ours. The novelist Marge Piercy in her memoir, *Sleeping with Cats*, understands why her cat Tamburlaine had run away from her: "She too wanted to be fully alive, to be free."

We humans sometimes express this need for freedom as a feeling of something missing in our lives. I wonder if that is how it manifests itself in confined cats, lions, or tigers in a zoo: a kind of nervousness, a pacing, and a longing for something they cannot identify. Rainer Maria Rilke, in his poem "The Panther," speaks of this animal "who feels faced by a thousand bars, and behind the bars, there is no world," because "his great will has been put to sleep." Who can ignore this when watching those magnificent large cats confined to their zoo cages, just pacing up and down with an intense expression—how can we not imagine them missing their freedom, even if they were born in captivity? It is how they evolved to live; of course they miss it. Perhaps that feeling of loss is the primary emotion of any caged or confined animal. It is often asked: How can animals (or humans, for that matter) miss what they have never known? I think it is possible. Obviously, we have evolved to take pleasure in some of our own capacities, and when these abilities are never used, or thwarted, there must be a concomitant sense of loss, or at the very least a sense of frustration.

When someone leaves her cat alone in her apartment all day while she goes to work, if she were to think about what the cat does with himself all day, she would almost certainly answer: He sleeps. Is it true? Cats do spend a lot of time sleeping (sixteen hours), but at their choosing, not because they are bored or are forced into inactivity. Her cat is probably sleeping when he would rather be playing or exploring or visiting exciting places. Like us.

For this reason, I am not in agreement with those who argue that a cat should never be let outdoors. I understand the impulse. The late Cleveland Amory, founder of the Fund for Animals, is adamant that cats, for their own safety, should be kept indoors. His friend Phil Maggitti gave him many good reasons, among them crossing the living room is far safer than the street; they will not contract rabies or be bitten by free-roaming animals; and their owners are less likely to have fleas, fungus, or worms than are free-roaming animals, so these owners will not transmit to them the feline leukemia virus, feline immunodeficiency virus, and other contagious diseases. He ends by stating that cats kept indoors will live longer, happier lives. It is true that they will probably live longer, but it does not follow that their lives will be happier. Cats evolved to stalk, to find secret hiding places, to pounce on leaves, to race up trees, to lie in the sun, to smell the natural smells of the natural world, to use their brilliantly evolved physical gifts to full capacity. The author Ida Mellen (in her book *The Science and Mystery of the Cat*) maintains that in cats the urge for a change and "the longing for the fields and woods, where they can smell the grass and flowers and see the breezes stirring the leaves, presumably is much the same as ours." I agree.

Cats can be happy without these things, but it is a peculiar kind of circumscribed happiness. Rather like us: we, too, confined to a room, could learn to cope, to even call ourselves happy, in a restricted sense of the word. That is not the fullest, richest sense of the word *happiness*. House arrest for cats or humans is, no matter how comfortable, a form of imprisonment. I have observed that the walks I take with the cats are not casual arrangements for them. They wait for them. At first, I was careful with the cats. I thought it was a fluke that they accompanied me and that I should feel honored, grateful, and never push them. Now, however, I know that they derive as much pleasure from the walks as I do and that no matter how many times a day I say, "Let's go for a walk," they are up for it and beat me to the door. Walking is definitely interesting to them. It improves the quality of their life. Maybe it does something else; maybe it has nostalgic value, in that it takes them back to a time in their ancestry when walking was the very essence of being a cat, and they are somehow reminded of this atavistic state.

Jeremy Angel, who studied zoology at Oxford, observed 120 stray cats in an animal refuge in eastern Hokkaido, in the northern part of the Japanese archipelago, for seven years. He takes the position that "the only really happy cat is a free cat, one that is able to wander out at will, to climb trees or step soundlessly through long grass, to hunt mice and rats, to sunbathe on roofs, to seek the solitude that its nature demands." He is correct. There is cat nature, as much as there is human nature, and happiness is intimately linked to the ability to give free expression to this nature. The absence of danger indoors may lead to a contented cat, but not necessarily a happy one, since some measure of happiness is achieved by overcoming dangers.

Is there compensation for being kept indoors all day? Can an indoor cat nonetheless be happy? Compared to what? is the only proper answer. A cat confined to a small apartment for the entire day can be happy. However, he pays a price. He may be anything but miserable, but it would be dishonest of us not to notice that he is missing out on much. If only we would realize that we cannot really own a cat, then we might begin to see how we have no right to confine a cat in this way without exhausting all alternatives (companion animals, cat doors, specially built indoor gardens, moving to the country; the list is long). People in cities must make an extra effort to find ways to enrich their cats' lives. The greatest enrichment, of course, is to give them a large dose of our company and affection.

On the other hand, I know that many people in large cities keep a single cat in a small house or apartment. The reason they have just one is that they believe the cat is solitary by nature and will be happier this way. This might be true if the cat has an entire valley at its disposal, but when it is confined to an apartment, and the people are gone all day, then I think there can be no argument but that the cat will be lonely. This feeling of loneliness may well be artificial, in the sense that we have imposed it upon the cat. In its natural habitat, probably no cat ever feels lonely. Our homes are not their natural environment, however. Cats will be far less lonely if they have a companion, another cat, or even a dog with whom they are on good terms, especially if the humans in their life are gone most of the time. If you are getting a kitten from a shelter, then it is a good idea to take two. Introduced young enough, they will almost certainly become close friends. There is no harm in introducing them to each other at the shelter to see how they like each other. Cats will definitely have preferences, and why not cater to them?

To say that indoor cats are deprived may be arrogance on

my part, as if we knew what was best for a cat. Enough people have told me of the rich lives their indoor cats lead and how much pleasure it brings both to the cat and to themselves, that it would be ridiculous for me to deny this and insist on my view that access to the outdoors is paramount in the life of a cat. Maybe I am just projecting my own great fear of confinement on animals who do not feel this. Perhaps it is a good thing that most cats left alone will spend the greater part of the time asleep—especially if when they awaken a beloved friend is there to play with them and fuss over them. I wish I could discuss this with a learned and loquacious cat! And it could be that the feeling of security and safety that such cats feel—in their attachment to the people they live with and their homes—compensates somewhat for the loss of freedom. To know every nook and cranny of their home, to know every hiding place, to be completely familiar with their surroundings, must give a certain pleasure and satisfaction.

In fact, there is an age-old controversy as to whether cats are more attached to humans or to places. The answer when it comes to dogs is very simple: Dogs want to be with humans, wherever those humans are. They are as happy in a tent as in a palace and do not seem to care how often "their" human changes homes or even countries. Cats, on the other hand, or so the conventional wisdom goes, become attached to a specific place and wish to return to that place with or without the people who lived there with them. Cats are obviously different from dogs, but I do not believe this accurately describes their feelings of attachment. I would argue that cats actually form a mental image, an internal map of their home *and* its inhabitants, and this map is precious to them. Disturbances to it are very upsetting to cats. They love familiarity; their survival depends on knowing what to expect. They must spend much of their inner life predicting, and they like to be right. A different home is terribly confusing, but so is a different person living in the same home. After all, cats are not indifferent when we leave. If Leila and I are gone for the afternoon, when we return the cats crowd around us and I can detect a hint of anxiety. "Where were you? Why did you leave?" Our going away for a week produces great anxiety, or perhaps it is confusion. The place is the same, but *we* are gone. In a sense, our home is an extension of us, for cats. Dogs are much more philosophical in this: what matters is the soul, not the body. All adventitious circumstances they are willing to overlook. Not so the cat. For cats, the envelope is part of the soul. There is no soul, for them, without its envelope. They are the ultimate materialists in this respect. For them, without our homes, we are like snails without their shells.

I always had difficulty understanding the attachment people felt to a country, probably because I have lived in so many of them. Now that I am living in the South Pacific, I sometimes become nostalgic for Berkeley, my original home and the place where I lived the longest in my life, the place I knew best. The terrain was familiar to me; I recognized the trees, the shrubs, and the grass. I felt comfortable there as I probably never will feel here. California is still "home" to me in a way that Auckland cannot be. Yet my deepest attachment is still to Leila, Ilan, and Manu. Where they are is home. So I understand dogs. Yet I am slowly becoming deeply attached to this beach, to the giant tree at the end of it, to my daily walks with the cats. I would not like to be separated from this place. The cats belong to this place as much as I do. So I understand the cats.

It makes sense from an evolutionary point of view that cats form deep attachments to place. After all, where do the emotions of a solitary species gravitate? Rarely to another animal. Therefore, by default there are only two possibilities left: the self and the external world—that is, the habitat, be it a den, a tree, or a territory. Cats are territorial; they seldom roam beyond the area that they consider their territory (slightly different from their home range, which could be defined as a smaller territory, more familiar and more intimate). This is partly for reasons of safety, since they know whatever dangers might lurk there and can be prepared for them. However, it may also be a function of their attachment to their specific territory—the comfortable familiarity that develops when a place becomes well-known after long acquaintance—lying in a favorite tree, warming themselves on a sunny windowsill, walking up a favorite path to the top of a hill. If the wild ancestors of our domestic cats, in their solitary state, felt few emotions for other cats (beyond the negative ones of anger, aggression, and fear), is it not possible that they poured their pent-up emotions into the landscape of their homeland? I am not suggesting that cats have the same aesthetic sense that we have, but they surely have pleasurable physical experiences much as we have and may have positive feelings about the landscapes associated with them.

In all fairness, it must be pointed out that some cats, probably more than we know about, choose to leave their homes and seek out another. This is likely more traumatic for the human who has been left than for the leaving cat. Either something about the former house displeases the cat or something about the new house pleases him more. Sometimes, though, this could simply be an example of the underlying independence of the cat coming to the fore. (Maybe, too, he is just curious or likes variety.) How, we wonder, could he simply pick himself up and leave, after all the affection I have lavished upon him? I can assure you, the cat feels no pity for your piteous condition. He is attuned to some other song, whose melody humans find it hard to hear. Sometimes the cat leaves a home not for a new one, but merely to live independently on his own. Something troubles us deeply about this decision. Having once known the joys of community, how can cats so easily revert to their ancestral solitariness? Do they not lose something precious? I suspect that the emotions that compensate for this loss are beyond our ken.

I confess that every time the cats come up the hill with me, I worry that they later might not return home. There are so many things that can happen to such small animals. Ilan and I walk to his school by following a small path that leads up the hill behind a neighbor's house. It goes through fairly thick bamboo groves, banana trees, and native subtropical rain forest. It is a very steep path, and for some reason the cats do not follow us. Possibly because they are so used to our usual route, which goes in the opposite direction, whenever we turn left instead of right, they stop and refuse to come. I am relieved, because when we reach the top, we come out on a main road, which we must cross to reach Ilan's school. I do not want the cats attempting to cross the road with us. Today it happened, though. When we went up the path, Miki and Minna Girl were right behind us. We told them to go back (foolishly speaking to them as if they were dogs), which instruction, of course, they totally ignored. I did not know what to do; Ilan was late for school, and I could not turn back to bring the cats home. Fortunately, they stopped at the street. All the way to the school, though, I worried that they would not find their way home. This is not reasonable, I realize, for cats are pretty much like us in this respect. Once they have been in a certain direction, they know their way back. I worried nonetheless that something could happen to them, and I was even more worried when a short time later, I returned to the path and they were not waiting for me. When I got home, though, Minna Girl greeted me with her chirp, and an hour later, Miki arrived, too. When I hugged him, he looked slightly disgusted. In his view, of course, there was no reason to make a fuss!

Sometimes a cat's attachment to a home or person can prove baffling. A South African friend, Ellie Ormond, who was working as a receptionist in a veterinary practice in Cape Town, told me of a woman who came in one day with an adorable female tortoiseshell cat. The woman said she needed to have the cat put down. She was having a dinner party that evening and did not want the cat to jump on the table! Ellie took the cat but of course did not put her down. Instead, she found her a home twelve miles away. Two weeks later, the horrified woman returned to the vet to complain that she had paid to have her cat put to sleep but had opened her front door that morning to find the cat sitting on the front doorsill, waiting to come inside. The cat sauntered in as if nothing had happened. Ellie thought the woman might have come to thank them for having made her see the error of her ways. Not at all. She wanted the deed done properly this time. Ellie found another good home with a friend living much farther away—and this time the cat stayed. Was the cat returning to the woman, or was the cat returning "home" in the broadest possible sense of that word, namely the place where the cat felt secure and least vulnerable? After all, it is not within a cat's abilities to know what nefarious plans are suppurating in the mind of a woman willing to kill her cat for a dinner party. The cat assumed, had to assume, that life in the home was safe. When she returned home, she was returning not necessarily to the woman, but to a way of life. She could not know that she had made a terrible mistake.

At least four and a half million domestic cats (not feral cats) enter shelters each year in the United States alone. Of these, almost four million are euthanized annually. It is a good thing cats cannot comprehend figures of this magnitude (can we, really?), to learn what is happening to their brothers and sisters across America. Figures for the worldwide killing of cats are simply unobtainable. Spaying and neutering my cats has, it is true, deprived me of any insight into feline sexuality. This is a small price to pay compared to the agony that is created by cats who are not neutered. Theoretically, a single mother cat and all her offspring, in seven years, could produce more than four hundred thousand cats! It is the height of irresponsibility not to neuter our cats.

I have often wondered whether cats in shelters slated for euthanasia know what is coming. I find it unbearable to walk through those parts of any shelter. The cats reach their paws through the cages, trying to attract your attention. If you stop to look, even an old cat, barely able to walk (for the older they are, the less likely they are to be adopted and the more likely to be killed, which is why no-kill shelters such as Brighthaven in Sonoma, California, are doing such a wonderful job by accepting old and ill cats whom nobody else wants) will often roll over and stretch out his paws in a pathetic attempt to make himself look cute and cuddly and desirable. So yes, perhaps they know that something is in store for them that is not good and could be prevented if only they could convince the next person who walks by to take them home, to give them life for just a little while longer. Even if they are oblivious to what is about to happen to them, perhaps they are so longing for companionship, for attachment, that they do what they can to make us want to rescue them.

There are model shelters, however. I visited one—Best Friends Animal Sanctuary, in southern Utah, just outside Kanab, situated on 350 acres in Angel Canyon, a red-rock canyon where the Anasazi left their petroglyphs. It is a no-kill shelter, where people can bring cats and other animals for life, with no fear of euthanasia. The cats live in huge indoor/outdoor atria, with up to fifty cats occupying each atrium. I was amazed at how well these cats got on with one another, how quickly they adjusted to this kind of group life. I asked one of the keepers, a cat lover (who else would volunteer for this kind of work?), whether the cats were happy. She assured me that they were, and they certainly looked that way. To what were they attached here? To the place, to other cats, or was it simply that they were attached to life and did not want to give up? As I watched them treading, ever so lightly, on this little piece of earth that was now their home, I could not help reflecting on how many things, by contrast, I was attached to and needed in order to feel happy.

If cats are able to feel attachment to places, to other cats, to us, if they can even love and feel compassion for other cats and human companions, why is it that somehow we so often feel inadequate in the presence of cats, a feeling that rarely overcomes humans in the company of dogs? (Somebody once quipped that only dogs have masters; cats have staff!) I think it is because cats reserve their most intense reactions for other animals—other cats, birds, even insects. When cats see these, they suddenly become intensely vigilant, interested, even obsessed. We humans are more like an afterthought for them: "Oh you, sure." We have not been relevant long enough in their evolutionary history to matter to them as much as the fly on the windowpane! We have forced them to live in a world where they have to pay attention to us, when perhaps they would really rather not. We are engaged in a futile effort to win their deepest loyalty, for in the end it exists where it belongs. Cats are true to themselves in a way rarely equaled by any other animal, human or otherwise.

FIVE

Jealousy

Minnalouche

There is no question that jealousy is an emotion with a great deal of cultural content. What is regarded as normal jealousy in one part of the world is considered pathological in another. The proper attitude of a husband to his wife in Arabic society, for example, is very different from what most Westerners consider warranted. We might label what he does as selfish possessiveness, and he might call us callously indifferent. So some jealousy is desirable in certain cultures, but not in others. If such difficulties exist among different social groups in humans, imagine how difficult it is to define jealousy in a different species. Nonetheless, we are certain (animal behaviorists have described it in the scientific literature about chimpanzees, orcas, and elephants) that it does exist among other animals and that it is especially evident in certain species, as anyone who has ever lived with a parrot knows (a parrot will sometimes not even permit a spouse access to his or her chosen person). All animals that pair for life, such as wolves and many birds, are especially susceptible to jealousy, probably because so much is invested. Cats do not pair for life, yet jealousy plays a role in their emotions.

A study carried out by psychologists at the University of Western Illinois found that 79 percent of cat owners reported signs of jealousy in their cats (a common one is sitting with their backs turned away from their owners). Few people doubt that cats feel jealousy. I too am sure cats become jealous, but I am not at all sure what "jealousy" means for cats. I am thinking of the everyday kind of jealousy that is so familiar to anyone whose house contains more than one cat. Moko, for one, seems to keep track of where I am in relation to the other cats, and especially to Megalamandira and Minnalouche, by whom he seems to feel greatly concerned. Notice how I hedge my words here. That is because I am not certain exactly what Moko is feeling (nor can I ask), only that something seems to concern him when I give attention to these two particular cats. What he feels appears to be something like jealousy, but it is not exactly like human jealousy, because the "rules" of human jealousy do not fit here. At night, Minnalouche likes to slip under the covers and roll her body up alongside mine. I know that whenever that happens, it will not be two minutes (usually less) before Moko, from wherever he has been lurking, has sensed (if that is what he does) that Minnalouche has gotten close to me and makes his appearance under the covers as well. There is no hissing or pushing, but she gets the message quickly enough because she immediately stops purring and within minutes is gone. I wish I could peek under the covers to see the look he gives her—or is his presence sufficiently eloquent? She does not fight about it and never seems cross either with him or with me.

Is this jealousy, or is it perhaps some feline rule of etiquette that Moko and Minnalouche both know and acknowledge and of which I remain ignorant? He is not unpleasant about it, just confident that Minnalouche has no place under the covers next to me. It is not that he wants the place for himself: shortly after she leaves, he leaves as well, satisfied of a job well done, I suppose. Cats belong to a different species, and it will always remain impossible for us to be absolutely certain what the experience of jealousy is like for a cat. Why did Moko seem to be jealous of Minnalouche only at night? He tolerated her during the day but drew the line at her sleeping in my bed. Have I missed some crucial detail of behavior that would make everything clear? It is easier for me to be sure that I understand the pleasure Moko and the other cats feel when they are leaping about ahead of me on the path to the top of the hill or rolling about in the sand on the beach. Is this because pleasure, somehow, is more easily translated across the species barrier than the more negatively tinged emotions?

It may be easier for cats, too, to read our positive emotions. When we were walking up the hill in the moonlight, I liked to imagine that the cats and I were all infected with the same sort of good feeling, that not only did I sense how happy they were, but they knew I was happy, too. Perhaps they even knew that what I felt was similar to their happiness. I do not, however, expect cats to be aware of human feelings of jealousy. Anger, of course, they understand—but only in the sense of immediate physical danger. (They seem not to take in—or is it that they just don't care—our annoyance at what they do or, even more, what they don't do in response to our requests.) Dogs have a proverbial terror of fighting between spouses, and I think cats like it none too much, either. Cats will often immediately leave a room where there is anything resembling a fight. I once had to shout at Ilan, who wanted to climb out on our roof, and I saw Minna's eyes grow large with wonder and fear. She ran away. Is this only because she felt threatened? Dogs evidently believe they must have done something wrong to have provoked the anger and perhaps fear punishment. Possibly cats simply do not like the noise, the commotion, and the threat to their regular ways. Both cats and dogs, though, are somewhat like children, taking the fight more seriously than the people having it.

I cannot imagine a cat jealous but who does not "know" (whatever that means for cats) she is jealous. Indeed, all the actions of a jealous cat are about this jealousy. She does not seek to hide it. Only humans seem to have feelings of which we, paradoxically, can remain unaware (it is a paradox because the very word *feeling* implies we are aware of it). I have seen people in a terrible mood, angry, sullen, raging about the house with no apparent reason. Until the reason becomes clear. To the person I mean. "Of course," the feeling dawns, "I was jealous. But I didn't even know it!" Knowing what we are feeling, giving it a name, is not trivial. Cats, however, do not need the intermediary of language. I am sure that Moko is never struck with the thought that his behavior toward Minnalouche is a product of jealousy, pure and simple. He does not wake up in the night suddenly understanding that he has been jealous of her from the day she arrived, with her silver spots and her flirtatious ways. He was never in any doubt as to how he felt and did not need to give it a name. He did not need to understand it; in fact, he needed only to feel it. And feel it he did!

When Megalamandira first came to live with us, Moko was also the one cat who refused from the beginning to allow him to become an integral part of our household. Yossie had no problems with him, even early on. Minnalouche was not thrilled, but she came around quickly. Miki and Megala soon began playing happily together. Moko, however, not only growled at Megala, he growled at anybody who had even been near Megala; he could not bear to smell his scent. Most people would call this pure jealousy. Megala was no threat except in Moko's mind; it is not as if we paid any less attention to Moko after Megala arrived. Moko was not getting any less love or petting or food from us because of Megala's presence. Megala was not affecting his life in any discernible way. The problem was all in Moko's head. Moko was prepared to have me share my love with three other cats, but he reached his limit with four. He bore no grudge against me; he was as friendly to me as ever (except when he first smelled Megala on my shirt and growled), but he could not be persuaded to be friendly to Megala. Though Megala was not the preoccupation of any of us, Moko was not having any of it. When I expressed my annoyance at Moko for behaving this way, by speaking to him in a stern tone of voice, he more or less ignored me. Mind you, Moko is never deferential. Rarely is any cat deferential to anyone, even another cat.

Of course, it is possible that what I saw as jealousy was simply Moko's way of protecting his turf. True, he had been willing to share it with three other cats, but he may have drawn the line there: "Not one more cat." But when resources are plentiful, as they are in our household, and a cat need not worry that he or she will receive less food, I cannot help thinking that the concern for the cat is the loss or diminution of affection rather than protecting territory.

There is a large Burmese cat who lives halfway up the hill behind our house who has recently taken to invading our house. He arrives and begins a high-pitched shrieking sound that brings all the cats running. They stare at him in horror, and Yossie, our biggest cat, begins to make a similar sound. I cannot imagine what this Burmese is seeking. Could he want to be friends, is he curious, does he need more food (unlikely, judging by his size), or is he spoiling for a fight? He leaves when I appear, so I do not know. Even if my cats are not territorial, this is their house, and they will not tolerate him in it. Almost no cat I have ever known is casual about a strange adult cat suddenly turning up in his or her house. How different they are with us. I am allowed to bring as many friends into the house as I like! Unlike parrots, cats do not resent faithfulness to other humans and are rarely jealous of people or even, it would seem, other animals. Just other cats. They resent, volubly, the presence of another cat.

The whole concept of dominance and territoriality in cats is in a state of flux. A number of studies of feral cat colonies in the last twenty years show so much variation that the dean of cat behavior, Paul Leyhausen, wrote recently that "in no case was the social structure of one group exactly like that of another. . . . Can all this variability have been produced by domestication? Or is there no social order in the species and all is but a reflection of environmental differences and pressures?" Defending a territory and deciding who is a more important cat has to do primarily with food and reproduction opportunities. Since most domestic cats today are spayed and neutered and have constant access to all the food they want, we do not see in everyday cat life much strife along these lines.

Cats have much less hierarchy than dogs and other canids, but they have some, and it is worth keeping in mind. The studies are contradictory, perhaps because hierarchy in cats is subtle, so you do not often see it, and it changes and depends on circumstances. It is not immutable, as with dogs. For example, a strange cat walking along a path of another cat's territory may well have precedence, simply because he got there first. Fights between cats are so dangerous for both sides—because they are usually evenly matched—that cats will do anything to avoid them, including not insisting on superior status. There is a certain amount of attempted bullying in our cat colony, but no one cat seems to always have the upper hand. There is no cause, then, for jealousy. They all sleep where they want, eat from one another's bowls, play with us, and walk along the beach in any order. Yossie plays more roughly than any of the others, and the consequence is merely that the other cats avoid him sometimes. Moko is the most athletic and will sometimes be carried away in play, so he too is occasionally avoided. Miki is always sought out, and I think this has to do with his egalitarian attitudes, pleasant to both humans and other cats.

Certainly the concept of subordinate and dominant plays little role in our colony. The cats do not fight for food or for my attention, and I have no sense that one cat is above any other. It is different in a feral cat colony. Howard Loxton reports that members of a colony of cats in Fitzroy Square, in central London, showed a peculiar deference to one small white cat, even though many of them were bigger, younger, and stronger cats. She was the matriarch—mother, grandmother, or aunt of most of the other cats. "She never showed any sign of aggression to the others. They simply gave way to her." So rank order in cats may be more similar to that in elephants (matriarchy) than to those in other animal species where large males rule despotically. Yet more evidence for the egalitarian streak among cats.

Humans often interpret mild jealousy as a necessary component of love. If a man admitted to feeling attracted to another woman, and his wife said it did not make the slightest difference to her, he might wonder whether she really loved him. Her motives, it is true, might be altruistic, valuing his happiness over her own, or impossibly sophisticated, but he would probably be suspicious nonetheless. (Jealousy in a cat can be seen as further evidence that cats are not thinking only about themselves. In its nonpathological form, it speaks of a kind of love.) We expect some form of jealousy to be part of the vocabulary of love. Exclusivity, however, is not something we seek to encourage in our cats. Many cats are one-person cats, expecting "their" human to be theirs and no other cat's.

Sometimes when I am lavishing attention on Minna, I will see Moko watching me. When he sees that I have seen him, he begins to groom himself (the classic displacement gesture for cats), as if he were indifferent to what Minna and I are doing. I am never sure. Maybe he is truly unconcerned, or maybe he is just trying to save face with me. Moko was not aggressive about his jealousy or even insistent. He gave few clues about just what it felt like for him to be jealous of Megala or

Minnalouche. Now, for some reason, it is all gone, history. At last, Moko is happy to share the bed with both of them! Often all five are together on our bed, and sometimes they are all together on their own, in a different room, or three of them are with Ilan, one with me, and one by himself. (Yossie rarely sleeps with anybody but himself.) Moko will now spend hours grooming Minnalouche and Megalamandira. Can we say his jealousy is spent? I think it is far more likely that Moko is no longer jealous of Megala because he has now become friends with him, and for some reason, perhaps peculiar to the syntactic rules of emotions in cats, a cat cannot feel friendship and jealousy for the same animal. Cat jealousy, then, would be less pathological than human jealousy, where jealousy crowds out the other emotions; cats seem to be able to have both. Human jealousy appears compatible with love, but love in cats seems to extinguish jealousy.

Unlike this mild, contained sort of jealousy, however, pathological jealousy seems to be very similar in cats and in people. The husband who demands to know his wife's every move, who suspects her constantly of infidelity, who cannot bear to let her out of his sight, flatters no woman. In fact, when a novel or a movie begins with a scene of this kind, we prepare ourselves to see a crime. Similarly, a cat who could not bear to see me pat one of the other cats, or who hissed when I kissed Leila, could not long remain a desirable member of our household. We expect a certain amount of restraint on the part of our cats, at least once they have been able to convince themselves that they are not missing something, that they will not have less of our love or affection if it is shared with the other cats.

I think cats are better at this kind of sharing than human beings because they come from litters that average between four and five. A litter of one is a rarity; often kittens have to share their mother with as many as ten siblings at once (although wild cats have smaller litters, no doubt regulated by the availability of food and other resources). One way they deal with this is to select a favorite nipple and then stick to it. Since each nipple (cats have eight) produces a different quality and quantity of milk, it has been suggested that the kittens compete for the best nipple. Kittens show remarkable fidelity to the nipple they have chosen and will rarely tolerate any substitution. I think this may account for some of the slightly eccentric choices that cats make later in life, such as refusing to go to bed except in a special spot or on a special blanket. These might almost be described as transitional objects, in the manner that psychiatrists talk about children choosing some object to represent the mother when she is absent (often, in fact, a blanket).

Cats have the advantage over humans that their seven-week period of nursing time with the mother is rarely interrupted. Traumatic separations are much less likely for kittens during the nursing period than for humans, whether due to illness on the part of either the mother or the infant or some problem the mother has nursing. If jealousy is the result of not having had enough body contact and enough nursing with the mother, then cats are at a great advantage over humans.

It is a myth that if you handle kittens early on, the mother will ignore them and refuse to nurse them. It is true, however, that if a mother feels you are interfering too much, she may take her kittens off and hide them. To be a truly socialized cat, one who is completely at ease with humans, it is important to be handled by people early (by four weeks) and as frequently as possible. A lack of contact with humans during the early weeks of a cat's life leads to problems later on. Michael Fox, a vice president of the Humane Society of the United States and a respected authority on cats, suggests in his book *Supercat* that kittens be handled from birth: "This entails simply picking the kitten up, repeatedly turning it around and upside down, and stroking it for a few minutes every day." A litter of kittens born where humans cannot reach them will hiss at people when handled at two or three weeks, whereas a different litter from the same mother, handled by humans daily, will be friendly.

In his account of the cats who lived at his farm in Maryland, the late Roger Caras spoke about how his family decided to make an entire litter into the most highly socialized cats of all time. The kittens, he explains, were held, stroked, carried, touched, and touched, and touched some more. It worked. They became completely people oriented, asking to be handled and wanting to spend all their time with humans. "It was simply a matter of intense, deliberate socialization. I am convinced it can be done with any domestic cat, if the holding and petting starts on its first day of life." This is still not the norm (I think it best to leave the mother alone for the first few days), which is one of the main reasons, I think, that we still find many cats who are skittish around humans (Moko, for example), in spite of a clear inclination to love us and be with us.

I believe that severe jealousy in a cat stems from very early disturbances in the mother/kitten relationship. Scientists who in studying deprivation in humans have removed kittens from their mothers at birth find they become deeply disturbed cats when (if!) they grow up. As in the infamous Harry Harlow studies of maternal deprivation of monkeys, we do not need laboratory experiments to tell us that monkeys, cats, and humans need the touch, nurture, and love that a mother brings to her children.

Sometimes when Moko is playing with a toy mouse, Megala will intervene and snatch it from him. Moko does not even get angry, let alone enraged; he does not feel possessive, as in "This toy is mine and mine alone." We have toys scattered throughout the house (both Ilan's and the cats'), and I have never seen the cats fight over a toy or insist on being the only one to use it. Ilan does this, both with other children and even occasionally with the cats themselves. They cede immediately. Cats do not have temper tantrums the way a two-year-old would, especially one whose toy was snatched away by another two-year-old. Everybody has seen the small boy who leaves aside a toy, only to begin wailing when another small boy then picks it up to play with. This possessiveness is found in all human cultures, but not, it would seem, in cat culture.

Most people who live with dogs have found themselves saying, "Stop being so greedy"—when the dog has eaten his fill but still guards his food bowl, for example. I have never had to say this to my cats. All five are content to eat out of a single bowl, if it is big enough and contains sufficient food. They do not fight over food, unless the food is alive. "Thou shalt not covet thy neighbor's catch" could well be a feline commandment, for we find that once a cat has caught prey and is carrying it about, no other cat, no matter how senior in rank, size, or age, is allowed near. The rule seems invariably respected. It is similar to the canine rule that once a dog has a bone in his mouth, no other dog may approach, even if the possessor is of the lowest rank. Even a puppy will growl a warning to a larger dog approaching, and the warning will be heeded. Cats, under these same circumstances, give a similar-sounding growl.

Cats possess nothing. Sometimes this strikes me with great force, as I worry about how to pay for our expensive house and everything it contains and then suddenly see my five cats, wearing nothing, owning nothing, and in some sense needing nothing that they are not born with, except food and shelter. When weaned at eight weeks old, a cat already has everything she is going to need in her life. (However, it takes females two years and male cats three years to reach their full weight.) Possessions do not possess cats and never define them the way they do humans. The desire to possess another person is the essence of jealousy.

Some academics believe that cats cannot experience true jealousy, because jealousy is a strictly human phenomenon. Jerome Neu, professor of philosophy at the University of California at Santa Cruz, says that jealousy is in its essence a set of thoughts and questions, doubts and fears, and not merely a sensation (feeling) or a transitory state like having a headache. I think this is wrong, because while human jealousy definitely has a strong cognitive component (that is, we think a great deal about what we are jealous of), it does have similarities with a headache (it appears suddenly and will not go away, whatever we think); it is most definitely a sensation. People can be completely overcome with the *feeling* of jealousy, even if they cannot give it a name or if they mistake its origin. Cat jealousy may be devoid of the cognitive element, but it is certainly not without a feeling component.

What I have never seen in cats is a particular kind of human jealousy for which there is the excellent German word *schadenfreude*, the pleasure we take in another person's failure, or suffering, or even destruction. Here we move into the area of envy, rather than jealousy strictly speaking. Envy is the feeling we have for wanting something that someone else possesses. When somebody who has something we want loses what he has, we may feel a strange and guilty satisfaction. If we cannot have what he has, at least neither can he any longer. Cats are not like this. Moko never smiles with satisfaction, metaphorically speaking, when Minna or Megala slips from a tree. In that sense, the jealousy that Moko feels is pure, and it is simple. He does not hate the other cats for the affection I give them; he simply wants something they have *at that moment*. The next moment he is their friend and bears them no ill will. Here is something we can learn from cats.

Envy, in humans, is considered—along with embarrassment and empathy—one of the self-conscious emotions, developed only in the second half of the second year of life, when children become self-aware. Nobody is certain that cats experience any of these emotions. There is nothing like blushing in cats to indicate embarrassment beyond the self-grooming after a fall, which may serve a very different function, as I have indicated elsewhere. Empathy is not always obvious or even evident in cats, though we have seen examples of cats comforting us when we are upset. That leaves envy. If we look at the ancestors of our domestic cats, they would not have had enough contact with others to envy them. This applies to embarrassment and empathy as well. The occasions for feeling these emotions in the normal life of the wild cat would have been rare. This does not mean, however, that cats are not self-aware, since we cannot apply the criteria we use in observing the development of such feelings in children to a completely different and nonsociable species. Whether cats are aware or not does not seem relevant to the range or intensity of their emotional lives.

Our son Manu El Natan was born on November 12, 2001. A few hours after he was born, we returned from the hospital to the house. I wondered how the five cats would react to the newest member of the household. Would they be wildly jealous? Was there a danger that Moko might hiss and attack, or would he accept Manu immediately as a member of the pride? I had many theories, but I was careful to put them aside for pure observation. What a disappointment! None of the cats took the slightest notice of him. We might as well have brought home a doll. Only when Manu began to cry did Minnalouche, who was sleeping on the bed at the time (so much for Leila's resolution that it would be too risky to let the cats into our room where he slept), suddenly become alert: she jumped up, ran to the source of the cry, took one look at Manu, and then, as if to say, "Oh, it's only him," immediately returned to the spot where she sleeps, put her head on her paws, and dozed on. Moko did not so much as glance at Manu. Miki came up and put his paw gently on his face, then walked away with no further evidence of curiosity. I believe that they were just not interested. After all, he is not like a child who can run after them and torment them or an adult who can play with them.

The fact that not one of them has shown the slightest hint of jealousy suggests to me that we need to rethink our notion of cat jealousy. I have heard of cats being annoyed, even furious, at the arrival of a new puppy in the house, but they do not seem to be jealous. That is reserved for other cats. We may be seeking jealousy in the wrong places; in fact, we may even be looking for the wrong thing altogether. We are bound by the human concept of jealousy and find it difficult, obviously, to think outside the confines of our own experience of jealousy. I can see why some scientists are so wary of anthropomorphism. Thinking like a human would lead us to expect a jealous fit in the cats upon the arrival of Manu. (Ilan told us on the first day he saw Manu that he knew a great name for him: the garbage boy.) I had anticipated a certain amount of jealousy of this being who was taking up so much of our energy, upon whom we clearly doted. To be honest, we were even scared about allowing the cats to see him for the first time. What if one of them leapt upon him in a jealous fit? What if they scratched him "accidentally" the way an older sibling might? We made a rule: Never allow the cats into the room unsupervised, and always keep the door to the bedroom closed at night so that the cats cannot enter and stealthily sit on his face or harm him in some way.

A surprising number of people believe that cats will smother and kill a baby out of jealousy. This is a double myth: there is no authenticated case of a cat killing a baby out of jealousy, or for any other reason, and no cat has ever smothered a child as far as I know (some tabloid newspaper accounts notwithstanding, and though an accidental death is not impossible). It parallels the myth of wolves killing and eating children (there is no documented case of an unprovoked attack on a human by any wolf in North America) and may stem from the fact that cats like to snuggle close to warm, small bodies with mother's milk on their breath. Slowly, all of the cats are beginning to take an interest in the baby, but it is not a hostile or a jealous interest. Miki loves to cuddle up to Manu now, but he is careful not to lie over his mouth. Does he know that would be dangerous? I cannot tell. Leila, the pediatrician, is not entirely convinced by my assurances and keeps a close eye on Miki. Megala came by this morning and sniffed every part of Manu's head—gingerly, to be sure, but with no hint of anything but benevolent curiosity. Now we leave the door open even when Manu is peacefully sleeping by himself and often find one of the cats curled up next to him on the bed. Sometimes there are two cats, but there is no evidence of jealousy, never any fighting for his attention.

• • •

Is jealousy more common in male cats than in females? My own experience would indicate that it is. Moko has been much more prone to jealous bouts than Minnalouche. There are good reasons for this. Domestic female cats are the only members of the cat family, besides lions, who engage in communal nursing. While some feral female cats live solitary lives, others form small groups with other females. They will often use a communal nest, and the females will take care of kittens not their own rather easily. Not only will they nurse them, they will also sever umbilical cords and carry the kittens to new nest sites if toms threaten them. While this behavior has been observed, notably by D. W. Macdonald from Oxford University, it has never been studied in any detail, so there is still much to be learned about this fascinating behavior. Since the nests are communal, it could be argued that the mothers are never certain which is their own kitten. (The kittens may know, since they have their own nipple, but not necessarily the mother.) This is unlikely, however, since the kittens are often born weeks apart, and in any event, a mother cat knows from smell who her kittens are. It seems to be a true cooperation. In an infanticidal species (which, according to Macdonald, cats are)—that is, where male cats seek reproductive advantage by killing the kittens of a different father (though what mechanism the tom uses for knowing a strange kitten, and whether this implies he recognizes his own kittens, is still unclear), the females often live together in this fashion, the better to protect their young. For while infanticide brings advantages to the male, there is nothing in it for the female. The females, moreover, are almost certainly related, usually a mother and her daughters from previous litters. Therefore, if female cats have had practice in sharing, or are genetically programmed to share, it would make sense that they are less jealous in general than male cats.

It is a quiet, brilliantly sunny Sunday afternoon. The ocean is that translucent blue it sometimes takes on. There are people swimming far out at sea; sailboats are passing by, moved by the slightest of breezes toward the bright green islands we see from our house. The seagulls are calling. All five cats are lying on the grass watching Ilan and his friends splashing in the gentle waves. They are lying close to one another without touching, but every once in a while one will roll on his back, stretch his arms out, and gently pull another cat over to him. I am sitting here, too, writing in my notebook. From time to time, one of the cats will get up and lazily stroll over to me for a pat and a cuddle. The others will look at us, blink their eyes in benevolent friendship, and go back to gazing out to sea. There is not a hint of jealousy. Is this the way it was always meant to be, for cats and for humans, when everybody has enough of what everyone wants—peace, sunshine, and love?

SIX

Fear

Megalamandira

Fear is so basic to the preservation from danger of every animal that even the most practical-minded scientist has no problem recognizing its pervasiveness in the emotional lives of all animals. Fear has even been observed in larve and embryos, such as larval African cichlid fish, which react to the slightest disturbance by suddenly stopping all movement and settling on the bottom of the ocean, or the five-day carrier pigeon embryo, which will jerk its head and neck away from a stimulus. Cats, too, seem to exhibit responses to pain before birth; studies confirm that a late fetus will show a distinct reaction to painful stimuli. Moreover, scientists report that if you pinch a kitten at birth, she will withdraw her limb and let out a cry of distress, indicating that even very young kittens experience fear (as in fear of pain).

Darwin was fascinated by how fear was physically manifested in cats. In *The Expression of the Emotions in Man and Animals*, he wrote, "Cats, when terrified, stand at full height, and arch their backs in a well-known and ridiculous fashion. They spit, hiss, or growl. The hair over the whole body, and especially on the tail, becomes erect." They are, as Darwin was possibly the first to recognize, attempting to make themselves look as big as possible, the better to face danger. We humans have a vestige of this: when we are frightened, our hair stands on end—a reminder of our earliest days when we too needed to look big and fierce in the hopes of warning off a predator.

Cats are notoriously susceptible to fear (much more so than, say, dogs), which is why children have come up with name-calling like " 'fraidy cat" and "scaredy cat." It is not entirely clear whether this fearfulness is something we project onto cats or whether they are really fearful and needed to be so as they evolved, having no friends to protect them. But the basic emotion that we observe in cats is perhaps not exactly fear, or even anxiety, but a state in some ways peculiar to cats: a kind of hypervigilance, an alertness to all possible sources of danger. This hypervigilant state exists in sharp contrast with their "contented" state. That two such opposite states can be both so essential to a cat's emotional identity shows how cats, perhaps more than any other animal except dogs, live in the moment. They can pass from one state to another like magicians.

Late at night, when the cats are on our bed, there will be a noise, and one cat will say to another: "I heard a noise, did you hear a noise?" The house is silent—I have heard nothing, but all of them prick up their ears and stare into the darkness. My heart starts to race, too. What are they hearing beyond my range? Then they settle, and I am convinced, absolutely, there is nothing to fear. I love watching their vigilance, the extremes of it: no noise, especially at night, when they are most alert, is beyond their keen interest, their need to investigate, to find the source and learn whether they are in any danger.

Cats hate loud noises, especially sudden loud noises, which may resemble to them the sound of a predator making a sudden lethal leap. I was holding Moko when Ilan let out a banshee yell and raced at me. Moko panicked, fled up my face, and left me with a bloody lip. I expected, and got, no apology from either cat or boy, both doing what comes naturally.

Some people say that nervousness is the very essence of a cat. I think it is true that all (or almost all) cats are nervous. They have to be, for they are such vulnerable creatures, small and entirely carnivorous—meaning they must hunt for a living, which exposes them to danger—and they cannot find safety in numbers the way so many other animals can. It is perhaps this nervousness that explains why cats rarely adjust easily to change. Change exposes them to new dangers. Cats like to know the way things are.

Moko has ear mites. Three times today I have tried to wrap him in a towel to hold him still while I give him his eardrops.

Three times he clawed his way out and gave me a look that says: "How can you be doing this to me? I thought you were my friend." The vet assures me that these mild drops are not painful, but it is a sensitive spot, and he simply cannot overcome the fear of having his ears touched. I would not call this fear irrational, but it is deep-seated and purely instinctive. He sees betrayal; I see help, and now the price I am paying for persisting is that he is becoming suspicious of me. I have given up on toweling him (our local vet, Martin Reid, is going to make a house call this afternoon—I hope my mere presence will not signal to Moko my continued treachery) and simply want to pet him again and tell him that I am indeed his friend, not his tormentor. He will have none of it; all my gestures in his direction are now suspect. Have I lost his trust permanently? I hope not. How unfair that cats sometimes cannot understand our true intentions.

Well, the vet has come and gone. It is wonderful to see a true professional at work. It took him only seconds to give Moko a needle, then clean out his ear. Moko hardly protested. I walked the vet back to his car, up the hill. Four of the five cats came along, of course. The vet was surprised. "I have never seen such harmonious cats going on an excursion like this," he said. "You have a genuine pride of cats here, something of which you can be very proud." I was pleased to hear this. Even better was the walk back: Moko approached me, and when I bent down he pushed his head against my hand, something he has not done for a while, since the ear infection began, in fact, a week ago. Clearly he was feeling much better, if not physically (although perhaps the relief from the injection and cleaning was instantaneous), then psychologically. Maybe this was his way of letting me know that he now understood I had been only trying to help him all along. He was in remarkable spirits and made a point of letting me touch his ears, something I have not been able to do for weeks now, obviously because they had been bothering him. His fear had abated.

Perhaps Moko was even feeling gratitude. It is often difficult to interpret gratitude, and we are anxious not to appear to be dupes and attribute to cats such a complex and humanlike emotion, but I have seen enough instances where I thought there really was no other way to understand what I was seeing. For animals who are susceptible to fear, the loss of what is after all an uncomfortable emotion may surely produce a feeling in them that is comparable to human gratitude. If they could speak, they might say they were grateful—though, being cats, they would certainly not overdo it.

When we reached the bottom of the hill, we encountered a man walking up from the beach with two dogs. The dogs saw the four cats and started to chase them. I was interested to see that all four cats stood their ground, raised their backs, and stared at the dogs. They were certainly frightened, but they were also confident. Was this only because I was with them?

Was it because they were four? Then they began a slow forward movement on stiff legs. The dogs stopped dead in their tracks, their tails drooped, and they began a retreat. They clearly were not about to take on four aroused cats. In fact, most dogs will not chase a cat who does not run. Running immediately triggers a dog's hunting instinct, especially when he is in a pack, and unless the cat can get away, the fight is unfair. If the cat stands its ground, however, it is a rare dog who will attack. Why is it that most cats have not learned this simple lesson? Is the instinctive fear of a larger, angry animal simply too deep-rooted for experience to gain the upper hand? It seems that my four cats have begun to sense their power, that refusing to run is to their advantage, and what force they have in numbers. I am curious to see whether they have really learned the lesson permanently or whether this was merely a fluke of the moment.

Most cats tend to be less than bold, if not downright timid. Miki is not, and his forwardness always astonishes people. A woman visiting the beach the other day watched Miki's calm, confident manner around strangers, dogs, and her children and told me that she felt she was meeting an important politician. She said it was a privilege to be around royalty! He does exude charisma—or is it just that he does things nobody expects? He also wanders unannounced into people's houses, exploring his kingdom.

When our little pride of four (sans Yossie, who resists acting as a pride member) and I set off up the hill, it is as if we are going to war: the four of them stare about, their ears pick up every possible sound. When we started going on these walks, the cats would accompany me only early in the morning or late in the evening, when there are few people about. Now they are bolder, and I rarely go up the hill without several of them following me. There are four dogs who live in adjacent houses on the beach, and the cats are suspicious of them—they watch them with eyes wide, tails twitching, poised for immediate flight—even though the dogs are no threat to them. Once the cats are certain that all is clear, they begin dancing their way along the path, moving sideways, all four large tails held high in the air. They look like a troupe of lemurs coming through the forest. It is, unquestionably, a lovely sight.

Are they frightened? Well, yes, a bit. They stop suddenly and then whirl around to see what is happening behind them. They stare at what I cannot see. Every noise, even the wind rustling the leaves on the trees, makes them jump. They dart off into the forest on both sides and race up trees. Of course, if they were really frightened, they would never have come with me in the first place. Nobody forced them to do so. Maybe they are just playing at being frightened or even just practicing the art of fear.

If I am gone for longer than an hour or two, when I return, usually only Miki is still waiting for me at the top of the hill. Sometimes, however, there is nobody. Then, as I begin walking down the path, I hear a long, plaintive call from inside the rain forest: one, two, three, or all four (only once five) of the cats are letting me know they have heard my footsteps and want me to identify myself. I call them—"Minnamikimoko!" and "Megalamandira!"—and in seconds all four of them rush out of the forest and present themselves. To be honest, I am not certain that their plaintive cry is a distress call, as scientists name it. It could be their idea of a joke: "You thought we were lost, haha, just fooling you!" Their sounds call attention to their presence, so I don't think they are crying out in fear. But they are always on the lookout for enemies, because this is how they evolved. It is probably not possible to overcome something as basic and hardwired as fear, even if constant experience teaches you there is nothing to be afraid of.

Cats may well share with humans the same basic roots of fear: a fear of annihilation, of grave bodily injury. There is nothing neurotic about such a fear in a world filled with automobiles, dogs, and other menacing objects. Given that this fear is legitimate, it seems natural that we should want to protect our cats from anything that would make them more vulnerable than they already are. That is why I am so opposed to people declawing their cats. I did not think I would need to address the issue in this book, because I assumed that everybody, from cat lovers to vets to academic animal behaviorists, would be aghast at the very mention of this operation. I was wrong. No less a cat lover than the late Roger Caras has a long section in his popular book *The Cats of Thistle Hill* where he defends the operation. It is, he says, a last-ditch effort to prevent a cat from being euthanized because the owner cannot stand the cat clawing the furniture and has tried everything to get the cat to stop. Rather than have him killed, Caras recommends declawing.

Of course, it is not nice to have a cat who claws the furniture. We just bought a new and expensive purple couch, and all five cats seem to think it is an extension of their scratching posts. I have placed the scratching posts at strategic spots all around the couch, and the cats weave their way past them to what they obviously see as the lovely purple giant of a scratching post. I yell no, and they stop for a moment with a puzzled look. They are surprised, genuinely surprised. "What's eating him?" Trying to say no to a cat is a futile exercise. It's like the proverbial "What part of 'no' do you not understand?" but here the answer is, "All of it." Of course, they are right. From their point of view, what could possibly be wrong with destroying a sofa? There are few sofas in a forest. That it bothers us does not seem to concern them very deeply. They don't like our being angry, for sure, but they do not seem to connect (except, perhaps, with great difficulty) our annoyance with their scratching. It just seems irrational to them, unfathomable. I have given up. If I were to redesign our house, I would look for a cat-friendly, not to mention child-friendly, architect, who would come up with brilliant solutions to save our furniture from shredding. We have begun draping large heavy fabrics over the couch, which helps somewhat. Ultimately, though, this is probably the price we pay for having a jungle animal inhabit our living room.

To contemplate a serious operation—requiring a thirty-six-hour hospital stay—that will deform the cat permanently just makes no sense to me. There is no benefit at all to the cats, zero. Declawing is done entirely for the convenience of the owner. I and others have noticed that the cats become permanently frightened—far more vigilant than normal—after the operation, because they have been disarmed; they are missing their most customary form of defense, sharp claws. A cat's abilities to walk, to climb, to mark (there are scent glands next to the claws that are activated by scratching), and to run are all affected, as is the cat's sense of balance.

Declawing is, after all, an amputation; it includes the removal of the terminal bone of the cat's toe, similar to the removal of the fingers of the human hand at the last knuckle, something, in fact, often done in torture. Declawing is dismemberment, plain and simple. Even cats who are confined to living indoors after the operation (which is imperative, since they are defenseless outside) undergo a gradual weakening of the muscles of the legs, shoulders, and back. Cats can no longer defend themselves, they can no longer climb trees, they can no longer sheathe and unsheathe their claws for the sheer pleasure of looking at their weapons; in short, they have been mutilated. I have no doubt that this operation will soon be illegal everywhere in the civilized world. It is already illegal in much of Western Europe, including the United Kingdom, Germany, Switzerland, Denmark, Sweden, the Netherlands, and Finland, as well as in Australia and New Zealand and many other countries. Of course, when Roger Caras says, "When it comes down to whether an animal loses its claws or its life, I opt for sustaining life," we are forced to agree with him. My point is that it should never come down to that. These cannot and must not be the only two alternatives.

My cats are slowly losing their fear of the neighborhood dogs. Two of the dogs are extremely large. Old Wagger, a black lab, has an enormous head. He is old, slow, and arthritic but still has a formidable voice and great presence. Ahi is part lab and part greyhound, perhaps, very tall, very muscular, and very good-natured. He is at least twenty times the size of Miki. However, as Miki approaches him now, the cat turns sideways, to make himself look bigger, and steps toward him on little mincing paws. Ahi backs away, and Miki stops to clean his claws, looking very proud of himself. Will they eventually learn to like each other?

Even Minna Girl, surprisingly, and Megala too have become curious about the dogs. It is more than curiosity; they have lost their fear and seem interested in becoming friends. Miki, especially, will walk straight up to any of the dogs who live on our beach and put his paw out to touch them. The problem is one of interpretation: none of the dogs is convinced of Miki's benign intentions, and each backs away. Miki continues to move toward them, his paw extended. They back away farther, convinced he means them harm. Maybe he does, but I think he merely wants a new friend.

Losing one's fear is exhilarating. Our walks through the rain forest as the cats become more familiar with it have become more exuberant. They race up trees and leap from branch to branch; they burrow under the leaves and jump into the giant silver-leafed ponga ferns. They have learned there is nothing to fear. That is an experience many people who move to New Zealand have; there are no snakes here, no poisonous insects except for one extremely shy spider, the katipo (rarely, if ever, seen), no poisonous plants. It is far different, in this respect, from its neighbor Australia, as well as the United States: there are no tarantulas, scorpions, rattlesnakes, or dangerous mammals of any kind (in fact, New Zealand has only one native mammal, a fruit bat!); its landscape is utterly benign. In the absence of predators, even the birds are friendlier than elsewhere; fantails will swoop down right next to you and follow you sometimes for a whole mile. They seem oblivious to the cats, who, alas, do not return the favor. It is not surprising that the cats are slowly losing their fear of their surroundings. There is nothing to be afraid of.

Cats are not native to New Zealand, one of the few places they are not. Other animals have not learned to fear them, so the cats take a terrible toll, I am afraid, on the native wildlife. Conservationists do not like cats here, and I cannot say I blame them. They are an introduced species, and a predatory one. Still, humans are an introduced species here as well and far more dangerous to native wildlife and to the ecological richness of the country. When I wrote a friend of mine, a professor of philosophy on the South Island, and asked him for his favorite cat story, he wrote back:

> Hey, Jeff, you got the wrong guy here. I'm a bird man. Bird people think of cats the way grain growers think of locusts. Even kittens, apotheoses of Cuteness, I perceive as Bird Killers Under Construction.

Birds here never learned to fear cats and are particularly vulnerable to an animal who approaches not to be friends but to get a meal. I do what I can to minimize the damage. I tried belling all five of the cats. The collars were found strewn around the beach and the rain forest within hours.

Minnalouche brings in benign hunting trophies: leaves. Her corner of the house, or at least her cache, the place she likes to use for herself, is slowly filling with all the leaves she brings in, a look of triumph on her small face as she deposits them on the floor and looks about fearfully to see if there is anyone around who might challenge her right to the kill. I cannot tell whether this is her idea of a joke or if she is serious. Cats are famous for bringing home their kill, often leaving them on pillows. As I remarked earlier, nobody knows for certain the significance. In the case of Minna's leaves, are these conscious displacements on her part? Could she possibly know that they are replacing the real kill? Is she just teasing us or even herself? It is a mysterious activity. I am delighted to have leaves and appalled to see small birds. As the cats grow, alas, their hunting skills improve along with their desire. My arrogance and ignorance is such that I was convinced I had weaned my cats entirely from the desire to hunt.

I thought I would try getting to the path at the end of the beach without the cats following me, so I suddenly began to jog along the beach. To my astonishment, both Miki and Moko began to jog alongside me. Miki was running, but Moko, with his thick, ringed tail held high in the air, was galloping—like a loping leopard, as a friend remarked. I wondered if he would continue if I turned around and jogged back. They were both right behind me. We did four laps of the beach, bringing the neighbors to their windows to stare at this bizarre sight. I have never heard of jogging cats before. How strange that two of my cats turn out to be dogs! The beach is becoming their territory now, a place that does not hold unknown terrors for them. When we walk up and down along the sand, it is almost as if they were patrolling it. Their step is more confident, their tails are held straighter. Their fear diminishes day by day. It is familiarity; I would love to believe it is also trust of me—that like a small lion pride, these cats look to me for guidance. If I let them know it is safe, they seem to accept my word for it. But this is not the case. It is not my word at all; it is their own experience. No cat will allow anything to substitute for direct experience. Cats come from Missouri: "Show me!" is their watchword.

I find it fascinating that my two jogging cats, Moko and Miki, can so easily overcome what must be an inborn fear of open places, where they are so vulnerable to forces they cannot control (in this case dogs), that they are willing to jog with me along the beach. Yossie, older (and perhaps wiser), watched with what I imagined to be disdain from the living room window. When I called him he turned his back (in disgust?) on the three of us and went to sleep. Minnalouche, however, wanted desperately to join us. She is so delicate and small, however, that she simply could not keep up. She made a brave attempt, trotting along, but soon withdrew and sat watching with interest. She is happy to come on the walks, but no running for her little body.

Unlike the other cats, Yossie has not mastered his fears. He is more cautious than the other four. Some people would maintain that he was born this way; I am convinced that his early experiences have made him this way. He must have been treated harshly. Unfortunately, if you give a cat reason to fear you, it will be very difficult to extinguish that fear, no matter how much your behavior toward the cat alters.

Evolutionary biologists maintain that humans have innate tendencies to fear certain animals (such as snakes) even when they have no experience of them, simply because these fears are of biological significance to the survival of the species. The problem is that many people have phobias for animals that represent absolutely no danger—butterflies, rabbits, lambs, even ladybirds and, alas, cats. Ailurophobia (from the Greek *ailuroi*, "tail wavers"—that is cats)—meaning not the hatred of cats, but the fear of them—is not at all uncommon. Some people are so upset by cats that if the word is mentioned, they turn red. Henry III fainted at the sight of a cat; others have gone into convulsions or suffered from temporary blindness, nausea, and lockjaw. It has been proposed in a classic article that preparedness to fear certain animals is a function not of the animal *per se*, but of their discrepancy from the human form. In other words, we fear animals who look least like us—insects, especially spiders, for example. It turns out that if somebody fears a specific animal, there is almost invariably some event in his or her early life to which the fear can be traced. If a person fears cats (as opposed to simply disliking them), we can be certain there has been some early and unfortunate experience with a cat. In and of itself, of course, a cat is not a terrifying animal. Some people, though, maintain that at one point in our history, large cats were dangerous to people, especially children. Even today, some lions and tigers (though always a minuscule proportion of them, less than 1 percent) will attack humans and regard them as prey. The striking resemblance of all members of the cat family would provide a seemingly rational reason for fearing cats. In every case where I have been able to ask about it, though, the fear originated in an early experience with an unfriendly cat.

Fear is one thing, dislike another. How about people who simply do not like cats? Everyone has noticed the strange behavior of cats when they are in the presence of somebody who does not like them. Often in a room full of cat fanatics, a cat will invariably seek out the one person who likes cats least. What is the reason for this? Is it a cat's idea of a good joke? Possibly, but there is a simpler explanation: People who do not like cats do not look at them. Cats appreciate this. They do not like to be stared at (which is what *they* do when they spot prey). If you blink when meeting a new cat, she will appreciate the courtesy. Looking away also helps. (Some human societies have similar rules; in Polynesian culture, one avoids eye contact with somebody of superior *mana.* Staring is universally considered at best rude and at worst an invitation to battle.) In addition, of course, people who hate cats not only do not look at them, they keep very still when they are noticed by cats, and this cats like as well. The less fidgeting the better. They have found the perfect lap. The big question, though, is whether the cat knows that the person they have terrified with their lap leap actively dislikes felines. Everyone has a different opinion on this matter. Most, though, admits cats could not care less!

Our neighbor Joan Chapple, New Zealand's first female plastic surgeon, does not like cats. At dinner the other night, I was saying that cats do not ever seem to be greedy. Leila wanted to know if they were greedy for affection. I thought there was no such thing. How could somebody want too much love? Joan intervened: "Oh yes, it's disgusting when they solicit affection by rubbing themselves against your legs. It gives me the creeps." Words of a person upon whom the feline charms are lost! I find nothing neurotic about the dislike of cats; I just find it inexplicable. But in fact, I have now learned that Joan's disapproval, like that of my friend the philosophy professor on the South Island, is based on the effect all imported predators are having on the naturally friendly New Zealand bird life. As Joan gets to know each of our cats, her general disapproval is becoming somewhat tempered by her regard for them as distinct individuals. Who, after all, could resist Miki forever? "Anywhere but in my garden," says Joan.

Do cats ever have phobias? Only, it would seem, of other cats. Cats fear all other (strange) cats but are not the least afraid of strange humans. (Of course, they fear all strange dogs as well.) It is odd that they have an almost instinctive trust in a foreign species (us) and the opposite instinct for their own kind. It is not the natural feline fearfulness that usually keeps cats from following us in the external world as we go about our business. If we encouraged them, if the environment were safe, they would spend much of their time with us by day as they often do by night. After all, if we learned our fear of cats from early experience, it follows that when cats fear humans they do so only because of early experiences as well.

I suspect no cat ever entirely relinquishes his or her self-reliance—at least not voluntarily. The fact that they must do so at visits to the veterinarian makes this an often traumatic event for all but the most placid of cats. I always feel that I am letting my cats down by taking them there. What they experience is pure fear, no matter how many times they have been, and I find taking my cats to the vet almost as unpleasant for me as it is for them. It is difficult for us to entirely understand this fear, because we feel that they must surely, by now, understand that it is for their benefit. Maybe they do and maybe they do not, but the fear does not seem to diminish. My dogs were the same. Humans also fear medical visits, but we resist, most of the time, the urge to run away just before they are due.

We are tempted to make the analogy with children. They too do not understand the purpose of doctor visits until they reach a certain age. It is no use explaining to a preverbal child that we mean them no harm in taking them for a vaccination, yet we insist in telling them this in any case. I know my cats do not understand what I am saying, but every time I take them to the vet, I attempt once again to justify my behavior. "Please," I say, "be reasonable. I do not want to do this. I have no choice. I am doing this for your own good." I know they do not understand, but I persist. I want them to understand. They do not, they will not, they cannot.

Yet many veterinarians and wildlife observers have reported that completely wild animals, and completely feral cats, will sometimes come to a human whom they would otherwise shun when they are suffering and need medical attention. The late James Herriot, the celebrated Scottish vet, gives a particularly convincing portrait of two such feral cats in his book *Cat*

Stories, and I have heard similar stories from many people. In the case described by Herriot, the two kittens were feral and had a tremendous fear of him and of his house. They would never, under any circumstances, enter the house or allow him to get near them, let alone touch them. Yet when they became ill with a severe case of feline rhinotracheitis virus, so ill that he was convinced they would die, they did enter his house, allow him to pick them up, examine them, and treat them. As he put it, "With a feeling that I was dreaming, I lifted each of them, limp and unresisting, and examined them." Once they were cured, they became completely wild again and would never again permit him to touch them. Whether they felt gratitude is something the good doctor does not allow himself to speculate about!

What are we to make of another common fear in cats, the fear of being in an automobile? Most cats dread this experience. The cats began congregating around our car at the top of the hill, and Miki, the boldest, would jump into the front seat when I opened the door. I assumed after a few weeks of this that they were ready for a ride. I was wrong. Moko howled as soon as I started to move. He was terrified, so I stopped immediately. Miki was not frightened, but I realized how dangerous a mobile cat could be in a car if he were to slip his body under the brake. As if that were not enough, as soon as we got to town and I opened the door, he made a dash for it. He could have been killed in the traffic. I decided never to take them again. Yet the mystery of the fear was still unsolved. Do they know how often their lives are taken by cars? Is it the unfamiliar movement? Why do dogs love to go for rides, their heads out the window, their ears blown back, sniffing in ecstasy, whereas even if a cat learns to tolerate driving, he or she is never ecstatic? (I knew one exception, my cat Omu, a Burmese who would sit by the door waiting for me to go out for a drive when I lived in Berkeley ten years ago. He was into the car in seconds and would sit happily at the front dashboard, watching the world go by.) Perhaps it is that they do not like to be displaced from their home unless they are the ones making the choice to move. It could be that somehow a car affronts their dignity of moving at their own pace, at their own will. If kittens were accustomed to cars from an early age, would they learn the pleasure of fast movement?

It seems that older cats have a greater tendency to fear than do kittens. It is a rare kitten that will not play with a total stranger, whereas most older cats are more circumspect, wary, even suspicious. They watch, they sniff, and then they approach. A kitten just hurls herself at us. Is it that as they mature they observe that there are more good reasons to fear the real world? Alternatively, is it that the kittens live in a never-never land that does not correspond to the real world?

"Experiments" (I hesitate to dignify them with the name) in social isolation of cats done in the 1950s by one Professor P. F. D. Seitz showed that cats who were isolated at two weeks of age until they were adults were "fearful" their entire adult lives (surprise, surprise). Thinking that we could learn about children from cats, a professor R. R. Collard a decade later "proved" that kittens not handled at all showed more fear of strangers than those handled by different people from five weeks until nine weeks of age. The question is whether a cat who has had a normal kittenhood might still develop into a naturally more fearful or at least cautious cat than a littermate.

Having something like a fearful temperament is a vexing question with humans and cats as well. Are some cats born with a naturally fearful temperament? Are there inborn differences? This is an ancient problem, one not to be solved here. We are increasingly inclined to believe that humans are born with very different temperaments, far more so than we did in the 1960s, when we believed more strongly in the influence of the environment. However, it is not as if there is any kind of closure. I remain unconvinced of the absolute ascendancy of nature over nurture. Two people from the same family can appear to come from different planets; similarly, two cats from the same litter are very different animals. Yet I would find it hard to believe that a cat who has been constantly loved, cuddled, and played with from the time he was very young and has suffered no traumas will be a diffident, unconfident, frightened cat. That just makes no sense to me. I would be glad to hear from people who have had a different experience, however. I have watched the inborn fearfulness and suspicion of cats grow steadily less over the time my cats have been with me. Every day their confidence increases, their sense of trust and mastery of their environment gets stronger. Today I watched them line up along the beach as I went swimming again. When I emerged they all came down to the water's edge to greet me. Their fear of the water itself is slowly ebbing. Will the day ever come when they will hurl themselves into the waves with me? Cats can swim instinctively, as can dogs and many other animals, even if they choose not to. Dogs like it, and few cats ever seem to. Can I somehow become their Roosevelt, exhorting them to fear nothing but fear itself? Would this not, for a cat, be pure deception? Safe as I try to make their world, the lives of cats are always dangerous, especially when they spend much of their time outdoors. I cannot always be there for them, and it goes against their very nature to trust a human with their lives so completely. I do not want to change the nature of my cats. Cats they should remain.

Because cats live in the moment, they fear only what is in front of them. Therefore cats fear fewer things than people do. Yi-Fu Tuan's *Landscapes of Fear* provides a long list of unreal things human beings have feared, including ghosts, witches, omens, curses, and gods. Cats, by contrast, are ultrarationalists. Their hyperalertness and their skittishness are based on real dangers. They fear only what they need to fear. The preternatural calm and self-possession we attribute to cats undoubtedly has something to do with the absence of irrational and imaginary fears. Worry about the past and fear of the future is unknown to them. Would that we could all be so free.

SEVEN

Anger

Miki, left, *and Moko*

There is no scarcity of books about human anger. Nearly five hundred publications have the word *anger* in the title, most of them books about how to control or eliminate it. Anger is felt to be an emotion far too animal-like for humans to accept. Yet almost nobody, beyond a few Buddhist elders, would claim to have transcended anger. Since we are not proud of anger, it is not surprising that it is, along with aggression and fear, one of the few emotions that no animal behavior scientist has any problem acknowledging being present in animals. In fact, no other emotions in animals have been studied as much as anger and aggression. Yet in the most authoritative book on cat behavior, John Bradshaw's *The Behaviour of the Domestic Cat*, the word *anger* does not figure in the index. Nor is it to be found in the otherwise complete feline encyclopedia by Desmond Morris, *Cat World*.

Evolutionary biologists recognize that anger, the roots of which are similar for cats and humans, is more important for humans than for cats. We need anger to keep us safe, to make us less vulnerable to attack. Anger energizes us, making us capable of defending ourselves with strength (often imaginary) and vigor. Ethologists say that in humans some forms of anger are more a liability than an asset and are therefore examples of cultural evolution outrunning biological evolution. An angry attack is almost never necessary in human societies. So we are always telling our children, "Use your words," not a statement one would expect to find translated into Cat. Nonetheless, we do expect some of the same civilized standards to apply to cats, and when they do not, we are not happy.

Cat anger is somewhat different from human anger, perhaps because it is so short-lived and so fixated upon the source of the anger. It is, unlike human anger, driven entirely by externals. I cannot imagine the ancestors of the cat, wandering by themselves in the African plains, feeling undirected anger. A generalized state of anger is not something we attribute to cats. Nor can we say that the cat feels angry at her prey. That is business, not feeling.

The expression "to get one's back up," about an angry human, derives from cat behavior—they arch their backs when they get angry. Anger in cats is hardly news. Still, it is far less prevalent than, say, anger in dogs. In the United States, dogs bite 4 million people every year, and in fact, every day some 914 people go to a hospital emergency room for treatment of dog bites. The vast majority of these bites are from dogs who have been maliciously trained to fight other dogs and have never felt kindness from a human. There are cat lovers who say cats never feel anger toward humans, only toward cats (and dogs). There is something to this, yet we must not overlook the fact that cats bite humans, too; in fact, experts estimate that cat bites are 1 to 15 percent of the figure quoted for dogs. That is still a substantial number of cat bites, and unlike dog bites, cats cause deep puncture wounds because of their long, thin, sharp teeth. What the statistics do not tell us is whether these bites come from cats who live in a home or from stray or feral cats. None of my five cats has ever bitten me, except for Yossie at the beginning of our relationship. My impression is that men are more often bitten by cats than are women, undoubtedly because they behave more aggressively around cats, and the cats feel they need to protect themselves. Moreover, more men than women seem to insist on "training" the cat or "teaching her a lesson" or just asserting their authority. They want to show the cat who is boss. Cats, however, are not interested in bosses or authority of any kind; it only raises their hackles. A man who insists, for example, on holding a cat after the cat has made it plain he or she wants to leave is going to anger the cat, and the next thing is a scratch or a bite.

Many men seem to expect obedience, a word for which there is no equivalent in any feline language. Discipline, control, submission, respect, deference—these are all concepts unknown to the world of cats. I suppose I must exclude lions and tigers, for they do seem to understand these terms or they would not be able to be trained to do things they normally dislike, such as leaping through hoops of fire (though it is probably best not to inquire too closely into how this is achieved). What exactly goes through their minds as they "obey" is hard to fathom. Whatever it is, the act of obedience, even for the big cats, is not natural, and I can see absolutely no justification for forcing these norms onto a species that has no use for them. Ordinary cats, in an ordinary household, will not take kindly to being disciplined in any way. "Live and let live" is very much a principal value in the feline world.

This is why punishment never works in training a cat. No cat psychologist would ever employ punishment or recommend it. Once you punish a cat, you will forever forfeit his trust. You will always be feared. Cleveland Amory hit his cat Snowball with a rolled-up newspaper. She gave him a withering look of disbelief, and he never did it again. A man I know rubbed his cat's nose in her urine. Bad mistake. She would not go near him for over a year. Why will a dog forgive you the next instant and a cat almost never? The dog expects to be punished; punishment is built into his genes. Mother dogs punish puppies from the beginning. A mother cat rarely punishes her young. A kitten's littermates will lash out when the playing gets too rough—indeed, this is how cats learn how far they can go when playing; but this is not punishment. Once kittens are weaned, they are independent hunters, cats already, and nobody tells them what to do or corrects their behavior. A grown cat probably confuses punishment with pure aggression.

When Yossie first came to us, he had the unnerving habit of stalking me. He would wait for me to walk by, then jump out and grab my legs. It did not feel like play to me, and the fact that he often clawed me was another clue that this was other than play. I began to fear that he saw me as a large rat and was out to dispatch me. My feelings (not to mention my ankles) were hurt, and I was getting angry with Yossie. I began asking people who knew about cats what to do. Everybody agreed the solution was simple: Yossie needed a companion. They were right. The minute our second cat came into the house, Yossie could turn his hunting instincts to play and be satisfied. This implies that Yossie was feeling not anger at me, merely frustration. I am not sure. After all, Yossie came from unknown circumstances and may well have had good reason to feel anger at humans. Once he was certain that I was not going to chuck him out, as had happened to him in the past, he might have felt comfortable letting me know that my kind were pretty awful sometimes. Nevertheless, it is interesting that he never stalked me again. Yossie is now a big and strong cat, and I find myself being cautious around him. When he has had enough of being stroked, I am quick to stop and get out of his way.

What makes a cat angry with humans? Generally, cats do not like us to infringe their dignity, and they do not like it when we do not respect their wishes. This is not, though, the stuff of cat-upon-cat anger. That is more mysterious.

Or is it? The one trait that has never been altered in any of the thirty-six species of wild cats (with the exception of lions and cheetahs) is solitary living. How does this essential trait translate to our domestic cats? For one thing, I think it helps explain why cat anger is short-lived: with the ancestral wild cats, there were few occasions for anger, for there were few meetings. A cat does not remain angry with another cat whom it hardly ever sees, and this fact seems to have left its mark on the cat's disinclination to feel angry over an extended period of time. What I most commonly see when the cats begin to play with one another is that shortly after the play starts to turn bad, when it does, I hear a sound that is a mixture of pain and rage, and one cat is chasing the other. What started out as play has turned into a fight. It happens all the time. There is not one aggressor to blame; all of the cats take their turn at chasing and being chased in this angry phase of the play. It definitely appears to no longer be play. So what has happened? It seems to me the cats have reached the limits of their ability to be sociable. It starts out with good intentions, but then their solitary nature takes over, and they want to be left alone and to be alone. On the other hand, if this interpretation is correct, I can then make no sense of the sleeping arrangements our cats choose, with the four of them at the foot of our bed or in a small cat bed that I have placed near the bed and sleeping in a tangle. That arrangement never turns ugly; one cat does not suddenly leap up and spit at another. Why the difference between sleeping and playing? Perhaps it is because play involves so many of the hunting attitudes and behaviors that it spills over into genuine predator/prey behavior.

I have observed, quite by accident, that we may be entirely mistaken in thinking the play has turned sour. Moko and Minna were playing when it suddenly took a different turn and they were locked in what looked like a nasty fight, with screeching and what I was sure was scratching that would lead to cuts. Somehow my hand slipped between the two cats, and I fully expected to end up with a badly mauled hand. To my surprise, what I felt was the batting of paws with claws sheathed. Everything had happened too quickly for the cats to have reacted to the sight of my hand by sheathing their claws; clearly the fight was never meant to be taken seriously by either cat. They may not have been playing, but they were definitely not out to hurt each other, either. This shows how easy it is for us to misinterpret a seemingly obvious scene from cat life.

Confinement, it seems to me, is a primary (if not *the* primary) source of anger toward humans on the part of cats. If you are holding a cat and he wants to be free, the more you hold him against his will, the angrier he will become. Even Minna Girl, the most friendly and gentle of cats, finds it intolerable. She looks pleadingly at me but wriggles more and more strongly. I have not tried to hold her for much longer than she wants, but I am convinced that after a point, even she would lash out at me. If we expand this notion a bit, we can see how being confined to a cage or a cat carrier would be torture to a cat. My rule for Ilan is that he never touch a cat with more than one hand, so that he remembers never to hold them against their will.

I have often wondered why the cats never become as angry with me as they do with one another. Yossie is wary of me and will not allow me to become too familiar physically, but none of them ever becomes visibly angry with me, any more than a kitten becomes angry with his or her mother—which could reinforce the idea that cats regard us as their mother. (Kittens rarely express anger or aggression toward their mothers. Whether this is true for adult cats, I am not certain, though my experience has been that it is. Female adult cats live in peace with their mothers all their life.) I think, though, that cats have a more complex notion of who we are.

Rage is rarely seen in cats, although Darwin saw it "well-exhibited by a savage cat whilst plagued by a boy." He does not make it clear whether the cat was savage by nature or by the circumstance, nor do we know how the boy plagued the cat, but we may assume the worst. Darwin goes on to suggest that the cat was *feeling* savage and was about to spring: "The animal assumes a crouching position, with the body extended; and the whole tail, or the tip alone, is lashed or curled from side to side. The hair is not in the least erect . . . the ears are closely pressed backwards; the mouth is partially opened, showing the teeth; the fore feet are occasionally struck out with protruded claws and the animal occasionally utters a fierce growl." In other words, the cat was not bluffing, and was not about to race away. It was preparing to fight back.

Other animals seem to understand the distinction that Darwin drew quite well, for when a cat assumes this position, most other animals back off. The large Labrador Ahi, who lives next door, often annoys the cats. When he has gone too far, Miki, Minna, and Megala will start walking toward him with stiff legs and a definite menace in their eyes. Ahi backs away as they continue to press forward. I have yet to see Ahi do anything but turn around and head for home. He is not about to be savaged by three determined cats. Since they are not pack animals, cats generally do not attack (or even hunt) in concert. Even individually they are formidable, and rare is the dog who will take on a cat once he is aroused in this fashion. The anger is potential rage; it is cool and deadly. It is as if the other animals know that once the cat has dispensed with tricks like putting his hair on end to make himself look bigger and more menacing, they are in even more danger. The cat has decided against flight and is ready to fight. I would love to know the outcome of the encounter Darwin described. I like to think that the boy who plagued the cat got his comeuppance and was unlikely to torture another.

With cats, the wagging tail is a clear physical signal that all is not well. I do not think that the cat is so much in the grip of anger that the lashing tail is merely its physical expression, the way trembling is; on the contrary, it is an honest advertisement of intent: "I am angry, and I intend to strike." Cats do this for the benefit of another cat, and they do it for our benefit as well, in the hopes, I believe, that we can read the signals and back off. As is the case with purring, I have never observed cats to lash their tails when they are alone. It is not just that they are feeling anger; it is anger they wish to convey. Tail lashing is a message meant for another living creature. You may not depend on the tail alone, however, for cats also lash their tail when they are playing, and there the signal is meant as play. We have to attend to the entire physiognomy of the animal if we wish to detect whether we are about to be played with or bitten. Interestingly, I have not seen the cats move their tails in this manner when they are about to pounce on prey. That act does not seem to cause them to feel anger (which is why we can say they are not being cruel when they hunt).

As in humans, anger in cats is closely related to fear and aggression. Typically, a scientist will claim that the rage displayed by a mother cat protecting her young seems like human rage only at a superficial level. In fact they are very different, because once the mother cat has frightened off the animal threatening her kittens, she will not pursue him. The mother cat is not interested in revenge. "Only humans," Richard and Bernice Lazarus claim with evident pride, "vow to avenge themselves against their enemies." The mother cat, in this example, is not angry, because the attack was not personal; it was to be expected. Her job is to deflect it and perhaps to be prepared for such in the future. She does not like the deed but is indifferent to the doer. Do we vow revenge on torrential rain? These are expected forces of nature to which we do not have a deep personal relation.

However, cats can bear grudges, which is why I am always telling Ilan not to chase the cats, not to throw sand at them, not to rush up to them with his arms flailing. "They will hate you," I explain. Cats will not tolerate such behavior from an adult (they seem to recognize the nature of childhood and make allowances), at least, and will forever avoid a person who is mean to them. They may not vow revenge, but they will not forgive or forget, as many people who live with cats know.

I have seen this occur with Moko, when his previous owner, Twink McCabe, came over to administer an herbal calming remedy after Minnalouche first arrived and Moko was so enraged at her presence. He was hissing, growling nonstop, stalking her continually, and we were in despair about what to do (in retrospect, I think had we simply waited another week or so, Moko would have calmed down on his own). Twink said she found it helpful to clamp clothespins on both sides of the neck of the cat: this made it easier to give the herbal medication orally; she said it never seemed to bother cats in the least. When she put the clothespins on Moko, however, he practically levitated, bounding high in the air, and then began screaming wildly, racing about the house, tearing at the pins. He looked like a completely savage animal. It was frightening. Even Twink said that in twenty years of working with cats she had never seen anything like this.

What was that rage all about? Moko was not in a good mood to begin with, because of Minna. The indignity of having something unknown stuck onto his skin (even though it caused no pain) was evidently too much and tipped him over the edge of annoyance into pure rage. Months later, when Twink came by to see all five of the cats, Moko took one look at her and ran out of the house. Nothing would persuade him to enter the house as long as she was there. The minute she left, he came back in and acted as if nothing had happened. He seemed to have total recall of the entire incident and could not bear to see the source of his misery, even if that misery was almost totally self-inflicted, for Twink had done nothing to harm him, at least not in our eyes (it is hard for us to see it from the cat's perspective).

One of the puzzles of cat behavior is the way even the most loving and trusted of cats will suddenly turn on you. You are stroking a belly, the cat is purring loudly, when in a flash the cat growls and grabs your hand with all four paws. As you try to extricate your hand, the cat puts more pressure and you realize you are caught. The only way out is to stay calm and speak soothingly to the cat, who might suddenly come to his or her senses and release you. "Phew, a narrow escape," is the usual thought. When cats have you in this tight grip, you realize how powerful they are, how sharp their claws are, and how badly they could injure you if they so chose. (It is amazing to consider how much damage a 4-pound cat can inflict on a 175-pound male human.) It almost feels like a warning: "Don't do that again." But do what? What did I do wrong? is the inevitable question. There is no obvious answer. It is similar to what cats do to other cats, when play might end up in a fight.

There have been many explanations for the sudden aggression, but none of them is completely convincing: The cat is embarrassed by his emotional dependence on you and is reasserting his independence; the cat is made to feel vulnerable by lying on her back and is regaining her composure; you have touched a delicate spot in the cat's anatomy; they do not want you to take them for granted; cats are unpredictable. Maybe all of these have a grain of truth in them, but I am not convinced we have uncovered the true explanation for this odd behavior. It is possible that the cat is torn between two opposing feelings. Veterinarians talk about a special form of aggression called "petting aggression," and the example they give is similar to the situation I have just described, when you are petting your cat, who is sitting on your lap and purring. Suddenly the cat bites your hand and jumps down. You cannot understand what you have done wrong. The explanation given by the veterinarian Bruce Fogle in his book *The Cat's Mind* is that while cats love to be touched by us, they are independent animals who seldom have natural physical contact with other animals except when fighting or mating. So the mind of the cat is in conflict. After all, in most circumstances, physical contact for the cat is a sign of danger.

Another possible explanation is that when two cats get ready to fight, the one who is playing the defensive will roll on his back and show all his paws with claws unsheathed, ready for action. If the aggressor continues to approach, the defender cat will clutch him and pull him toward his open jaws, while his hind feet trample and rake his exposed underside. This act is called the "rollover self-defense" and has been described for many cat species, not just domestic cats. Roger Caras says that when cats are in great danger, they will roll over on their backs and rake with their hind claws. This often inhibits an aggressor's attack, because the physical posture is so effective. It is a defensive position that is easily turned into an offensive one. Could it not be, then, that we somehow trigger this defensive/offensive rollover by touching the exposed stomach? This position is very different from the same one taken by a dog, which is a signal of defeat. The dog is effectively helpless when he is exposed in this manner, and by assuming this position, he is signaling to the other dog (or to us) that he has given up the fight. Cats, on the other hand, equally matched in lethal equipment, are always ready to defend themselves effectively.

With a human hand and arm captured in this way, the cat can tear with his hind feet against our skin and do serious damage. The point not entirely answered, though, is why play and the obvious pleasure the purring cat takes in our stroking her belly should suddenly trigger a completely different scenario in her mind, causing her to behave as if attacked. It feels like a delusion, or even a short-lived minipsychosis. The cat, after all, is not in great danger; why behave as if she were? I do not think it is play, for even if they rarely break the skin with a bite, one is aware that matters could escalate in no time.

Miki and Megala do not do it. Yossie does it so consistently that nobody wants to pat him when he is on his back. Moko will never go on his back for a person (though he loves to roll in the sand like a dog). That leaves Minnalouche: she starts to grab your hand and bite, but then a strange look will come over her face, as if she is remembering something about her love for you, and she stops. She is never angry and never aggressive with us. Perhaps the belly is simply a very sensitive part of cats, and they do not really like having it touched, even by somebody they trust and even love.

Most psychologists do not regard anger as a basic emotion but instead want to break it down into certain primary fears, such as the fear of being separated from the person one loves or of losing love altogether. The dominant paradigm is by the late

John Bowlby (1907–1990), a prominent psychiatrist heavily influenced by ethology and the author of attachment theory. His views are to be found in his classic trilogy, *Attachment and Loss*. For Bowlby, most anger is a defense against separation, and while he did not originate the term *separation anxiety* (as early as 1905, Freud explained a child's anxiety in the dark as being due to "the absence of someone he loved"), he certainly gave it the most attention in his three volumes. For Bowlby, "a child's pleasure in his mother's presence" is as primary as "his pleasure in food and warmth." A particularly poignant example is described by Bowlby of Laura, a child of two years and four months who was filmed in a location away from home while her mother had to stay in the hospital for eight days for an operation. When the film was shown to her parents, Laura inadvertently woke up and came in for the last segment, which showed her delight at the prospect of returning home when her shoes were produced. The lights went up and Laura turned away from her mother, to be picked up by her father, and then said to her mother reproachfully, "Where *was* you, Mummy? Where *was* you?"

Cats experience something very similar. They get angry at being left. Most people who live with cats know that when they leave their cats and return, the cats express anger. Sometimes they stop using the litter box, choosing the bed instead; sometimes they bite. They can turn their backs on you or pointedly ignore you. One could argue that from a theoretical view, the cat who has been separated too early from her mother or has been traumatized in early life by being left by humans will have the greatest anger. I think there is something to this. Of our five cats, only Yossie's childhood is something of a mystery. It must have contained trauma or he would not have turned up at somebody's house a year ago, insistent on being taken in. He may originally have had a home, but for some time he did not, and this early experience has marked him, and marked him with anger. Yossie is the angriest of our cats and also the one least likely to inhibit his anger.

When children come to play with Ilan, and they are all on the beach, they invariably pick up whichever cat is nearest and carry him about. The first time I saw a child picking up Megala, I came running to warn her of the danger. To my surprise, Megala had gone limp and was not objecting in the slightest to being carried. Megala was not enjoying the experience, but he was not panicked, either. He seemed to have made the perfectly valid distinction between a child and me. The child would not be able to interpret his anger and would not understand it. Therefore Megala did not display it. If this is not intelligent insight, I do not know what it is.

While it is true that a cat can become angry—in fact, does so easily and frequently—the same cat is perfectly able to inhibit his or her anger. We often speak of people who are "overcome" by anger, implying that the anger builds up in them (the old Freudian hydraulic model) and then demands release. Cats are different. They are not overcome by anger, unless something in the environment threatens their bodily integrity in a serious fashion. Cats, unlike children (and the great apes), do not have temper tantrums. A temper tantrum, of course, is always directed at someone else. Children do not have temper tantrums when they are by themselves. Because cats have been self-reliant, and are still largely so, their anger ("annoyance" is probably the better word here) is healthier, directed at a situation they want to change. This is also why you will rarely see inner-directed anger in a cat. At least I have never seen it. I have heard of cats chewing their tails in a pathological manner, and I suppose one could claim this was displaced anger, but I am not sure this concept applies to cats.

Is there a difference between male and female cats when it comes to anger? My experience with many cats over the years has been consistent: Females are far less prone to anger than are males. We now have four males, with Minnalouche as the lone female. She is definitely the mildest of all the five cats, the least likely to show any anger or aggression. Human research clearly shows that women have more intense reactions across a range of emotions, including annoyance, disgust, sadness, warmth, happiness, hurt, fear, and nervousness. The conclusion is that women are better than men both at recognizing feelings in others and at verbally and facially expressing a wide variety of feelings both in words and in their facial expressions. The exception is anger. We have a great body of literature indicating the importance of familial socialization influences, as well as language and culture. Yet if we find that these same generalizations apply to female as opposed to male cats, then it is less likely that it is purely a cultural phenomenon. In my experience, female cats *do* tend to be less aggressive than male cats, more affectionate, or at least more expressive of affection, and less quick to anger. Judging from purely anecdotal sources, many other people who live with cats have had similar experiences. No research has ever been carried out on this, but I would speculate that the reason for this difference could be childbearing and child rearing. Cats are wonderful mothers, protective, affectionate, and patient. Male cats, with few exceptions, never participate in child rearing beyond protecting (and even that sporadically) the female. Because mothers are more used to sharing, to giving their time, energy, care, and love to others, they get less angry at demands that would drive a male cat (or human) crazy.

There is a new movement in the study of emotions, known as the "social constructionist view," which maintains that cultural and linguistic factors are more important than biology. ("All human emotions are social products, based on beliefs, shaped by language, and derived from culture. They are not modifications of natural states.") The music critic H. T. Finck was the first to advance this view of emotion, when he argued that not everyone has the talent for love. Like music, he said, it must be cultivated. He argued that love, as we know it in the West, is of rather recent origin, dating to 1274 when Dante fell in love with Beatrice! This theory suggests that we are deluded to believe that when we are experiencing anger, we are being caused to have it, which explains the title of an influential article by the philosopher C. Terry Warner—"Anger and Similar Delusions." According to Warner, the other emotions that share this property are contempt, hate, embarrassment, dread, jealousy, self-pity, boredom, and many, but not all, instances of what we call indignation, anxiety, guilt, and indifference. If we apply this theory to a cat's anger, however, we can see where it must fail. There is no delusion in the cat's anger. The cat may be wrong in directing his anger at us; we may have done nothing to merit it, but he would not be angry were it not for something that happened to him "out there" in the external world. Anger is not all in the cat's head.

Cats are a good testing ground for this hypothesis because they are so often alone and in their ancestral form were almost entirely alone. The entire vexed question of what emotions a solitary animal feels is difficult to grasp, let alone to answer. It is especially difficult for us, a sociable species, to understand what an emotion might feel like divorced from other people. You need not be a social constructionist to see the truth in claiming that almost all our emotions are other directed. A solitary wild cat, however, rarely has an object upon whom to project his feelings, whether the feeling is anger, love, or many of the other emotions we have been considering in this book. Nobody has ever watched a solitary wild cat in his original habitat long enough to even guess at what emotions are coursing through his mind. Anger, however, is not likely to be uppermost. Even if he, say, falls off a branch or jumps short in attacking a prey, he is not likely to feel anger either at the prey or at himself. Self-directed anger, which could well have a useful purpose (to determine to do better next time), strikes me as highly improbable in a solitary cat, because anger seems to be social in origin.

Apart from the raw emotion provoked in a mother wild cat protecting her kittens, anger and rage would likely have developed with the domestication of the cat, brought into our world. Have we enriched her emotional life or impoverished it? It is difficult to say, but one thing is certain: We have complicated it. In fact, we might reinterpret cat anger and the edginess that many cats display in human interactions as a recognition that we are somehow disturbing the natural life of these solitary felines for our own purpose—one that no doubt confers benefits on them, but not one they would themselves have chosen if given the chance. In this understanding, cat anger is justified indignation at a transformation that was never willed, only endured, sometimes with grace, at other times with snarls and surliness, whose origins it would behoove us to understand.

EIGHT

Curiosity

Megalamandira, left, *and Moko*

Curiosity is actually a rather, well, curious emotion—if it can be called an emotion at all. I call it an emotion since we can *feel* curiosity, and it certainly seems to be an essential emotion in cats, if not their natural state. The purpose of curiosity in the evolution of any species is obvious: How can it not be useful to gather information about potential friends and enemies, for example?

Several years ago, I saw video footage of wildebeest in Africa running from a lion and then returning "out of pure curiosity," said the narrator. It did indeed seem that way. However, why would a prey animal be curious about a predator, when such curiosity could cost him his life? There are many theories, most focusing on the animal's need to conduct research, to gather information about the enemy. Who is he? How strong is he? How fast? How deadly? We may not be the enemy to our cats, but they seem to be researching us, too. They watch us very carefully, study us, and are ever vigilant.

When I return from a trip into town, my cats are waiting to smell my shoes. They know I have been somewhere that is of interest to them, perhaps even near another cat. When I come home with a package, all five cats generally mob me to see what it is I am bringing home. It is not that they hope for some new treat for themselves; they are no more interested in what I buy for them than in what I buy for myself (very different from Ilan in this respect); they appear genuinely curious *for its own sake*. When we go for a walk, they seem curious to see what I am curious about, and when I stop to examine a bush or a flower, they crowd in to see what I see. Even when I type on the computer, they want to know what I am doing, and invariably Minnalouche will lie down on the keyboard and begin to purr. Miki watches the cursor intently and paws it. I would love for them to know I am writing about them, but of course they do not and cannot. If they did, would they care? Almost certainly not.

If you bring a strange object into your house, your dog will sniff it briefly and quickly lose interest. Your cat, on the other hand, can become completely absorbed. If it is an object she can enter, such as a box or a bag, she will do so immediately. At times cats appear heedless of danger, so intent are they in exploring the new, and undoubtedly this is where the expression "Curiosity killed the cat" came from.

It is true that when my cats are driven by curiosity, it evidently overrides their sense of caution. We have a deck off our bedroom; one side has a very narrow balcony that opens onto a twenty-foot drop to a cement patio, so we have put up a large glass partition to make certain no child ever approaches the edge. Before it went up, the cats were used to walking along the edge onto the roof on the other side. I always held my breath when they did this, for one false step and they would fall to certain harm. Once the glass was up, the ledge was even smaller, and I was certain the cats would never attempt to maneuver along it again. I was wrong. They squeeze their bodies through a small opening in the glass and then walk along the edge of the roof, just as they always had. It is as if they are driven there by pure curiosity; after all, there is no place they need to get to that they could not reach more easily in a safer way. They show no fear whatever. How can they not? It is dangerous, even for a cat. Do they evaluate the situation differently from me? Are they just foolhardy? All of them (except Yossie, who is too big) do it, and do it frequently, and I never cease to gasp when they do. Cats have the ability to fall from great heights unscathed, but even if they could survive such a fall, surely they would not endanger themselves for no reason. It puzzles me.

Proverbially, cats have nine lives because they are always in danger of losing one, killed by curiosity. When I was a child, we had a retort to "Curiosity killed the cat," which was that satisfaction (in learning the answer) brought him back, giving a neat explanation for the seeming indestructibility of cats. In fact, by learning what they need to know about us and about our world, cats indeed often save their lives. Maybe cats normally avoid water, for example, because they know how easy it would be for them to drown. On the other hand, when Miki leaps into our kayak for a ride, he is not thinking about the danger; his curiosity has gotten the better of him, or his need for a thrill has taken over.

Cats have acute senses that serve their desire to investigate. I have mentioned that when the cats are sleeping with me, they will often prick up their ears at sounds too faint for me to hear. I trust their senses over my own, because cats hear far better than we do. Since they need to hear birds and rodent sounds at very high frequencies, cats respond to tones up to 60 kHz. Rodents squeak in the 20–50 kHz range. Humans hear between 15 and 20 kHz (dogs go up to 35, better than us but not as good as a cat).

The eyes of cats sometimes give the impression of surprise or intense curiosity. This is because cats have the largest eyes of all domestic animals, relative to their size. They have big pupils that let in more light at night. Cats require only one-sixth the amount of light that we need to see. This makes them superb nocturnal hunters. Especially at twilight or on a starry night, cats see as well as we do during bright daylight. There is no sense that the day is ending for a cat. At night, it just begins. Since we tend to play with cats during the day, they have two days for our one. Their sixteen hours of sleep are generally taken in tiny increments, catnaps, which is why we have that expression.

During bright sunlight, the cat needs even more protection than dark glasses afford us, which is why their pupils narrow to slits; they need this extremely fine control to avoid dazzle. As they emerge into sunlight, we see our cats blink and then adjust their pupils. Cats like to lie in the sun, but invariably their eyes are closed.

The eyes of all cats have a reflective quality. They shine when a flashlight is aimed at them at night. A crystal-like mirror in their eyes, known as the *tapetum lucidum*, causes the reflection; it allows cats a second chance to take in something they might have missed in dim light. It lines the back of the retina, acting as a mirror, reflecting light unabsorbed the first time it passed through the retina. The glow, known as "eye shine," is what we see when light strikes a cat's eyes in a dark room. To medieval witch-hunters, this reflective quality gave the cat a demonic look, perhaps because the color of the light reflected back depends on the color of the cat's eyes. Cats with green or yellow eyes tend to reflect greenish light. Cats with blue eyes, such as Siamese, tend to reflect reddish light.

As a night hunter, the cat does not require color the way humans do or a bird searching for brightly colored fruit would. Researchers believe, based on behavioral and physiological studies, that the primary color a domestic cat can see is green, with some blue. Wild cats that hunt in the bright midday sun, such as the cheetah and the Spanish wild cat, have two or three times as many cones in the slit of their eyes as the domestic cat and are able to see color fully. Interestingly, it has recently been discovered that at birth domestic cats have a sophisticated (although only latent) aptitude for real color vision. They have no use for it, so it disappears. Has it been retained in case they ever have to return to their original lifestyle? Although it appears nobody knows for certain, feral cats seem to see more colors than domestic cats. Should a domestic cat turn feral, would its color vision improve? Cats see the world more or less the way we do at twilight, when the landscape is drained of color saturation. They lose the color but have traded it for sharpness, since this was more important to them as nocturnal hunters. Cats are active at night because they can be. In time, however, most domestic cats adjust to our schedule and sleep when we do, simply because they appear to want us to witness their activities.

Curiosity seems to have a pure pleasure component; it is fun to see something new, and cats appear to be no more immune to this than we are. Why should curiosity for its own sake be confined to our own species?

Some people have claimed that cats never waste their feelings and are therefore curious only about what concerns them directly. I do not believe it. Ilan, at five, has just learned to ride a bike without training wheels. Four of the cats were lined up on the little path in front of our house, watching the proceedings with the utmost attention. As Ilan rode past, shouting in jubilation, "I am riding, I am riding!" the cats watched him intently, following his progress all the way along the path. This seemed like pure curiosity, unrelated to anything connected to the lives of the cats. What have they, after all, to do with a bicycle? The cats soon lost interest in Ilan's bike riding. However, when later that month I took out my own bike for the first time from the shed, the cats ran up to me, pushing between my legs, trying to jump up on the seat, smelling and staring at the bike, then looking up at me: "What is this for?"

Recently we bought a sea kayak and took it down to the beach. All the cats, of course, followed us and watched with what looked like fascination as Ilan got in and paddled off. Some of the neighbors were watching as well, since it is not every day that you see a five-year-old get into a boat and calmly paddle away. (We were a bit nervous, too, but he had a life vest, he had taken lessons, and it was a beautiful transparent flat ocean at low tide, so shallow that one could walk far out to sea.) The cats were not aware of the novelty of it in this sense but seemed intrigued merely by the idea of a bright plastic thing going into the water with one of us in it; they looked at us, they looked back at the boat, they looked out to sea. Clearly they could not fathom why anyone would do this. They all step into the boat when it is on shore, but I am reluctant to simply shove off with them on it because they've never actually gone in the water of their own accord, and despite their curiosity, it might be a terrifying experience for them. I like the way they look, though, with their bright coats on this red boat in a turquoise sea. When we are not around, the cats seem to have no interest in the kayak, though they love an old boat turned upside down on the beach, because they can get under it and chase the next unsuspecting cat who approaches.

A month later: Miki (of course it had to be Miki) simply couldn't stand it any longer, and yesterday he jumped into the kayak just as Ilan was pulling out. He stood proud and tall on the prow, like a painted wooden figurehead. Two neighbors came out to take photos of Miki on his first kayak adventure. As Ilan headed back toward shore, Miki took a great leap and landed safely on the other side of the small breaking waves. He looked very pleased with himself. It struck me that what we were seeing was unadulterated curiosity, unrelated to any function it might serve in Miki's life. There was no obvious purpose to his little kayak trip; it was driven entirely by pleasure and curiosity. The curiosity of cats is, like their affection, of a purity and intensity rarely seen in humans. We would be jaded when faced with the fiftieth paper bag. Not so our cats.

One of the cats shows an equal fascination with gardening. Our roof consists of native ice plants, and I weed them once a week. Moko is enchanted; he loves to watch me pull up the weeds as he attempts to hide underneath the fat little kernels of the ice plants, trying to dig himself under them. He watches my every move and seems disappointed when the work is over. Is it that he loves to see what will fly up and away as I weed, the little crickets and especially the bright green praying mantises that I need to protect from him? Whether there are bugs or not, he is equally curious about the activity itself. I cannot see any utilitarian purpose to this curiosity; however, it seems essential to him.

Swimming is not one of the natural abilities of cats (though it appears that all animals can swim if they have to). So curious are Minna, Miki, Moko, and Megala about my love for the water that it seems to have rubbed off on them. Not only do they watch me swim, but Minna now lies just at the edge of the waves and allows them to lap up against her side. She stares with pure curiosity at the water touching her fur. Given the cats' interest, had I started training them early enough, would it have been possible to teach them to come with me into the ocean? Imagine it: First we go for a walk in the subtropical rain forest, then we jog, then we all go in for a swim together!

The novelist Ian MacMillan, who lives in Hawaii, told me that he had a gray tiger cat named Fred who liked to go swimming with him and his family on Diamond Head beach in 1969. The first time it happened, she had been a little kitten and simply followed them into the water. After that, it became habitual: when they went into the water, she would go, too. He could not honestly tell, though, whether she went out of desperation to be with the family or for the pure enjoyment of it. Is it possible she was just curious as to how it felt?

If curiosity is the natural state of a healthy cat, then boredom could be considered the equivalent of an illness. Boredom in cats, though, is difficult for humans to ascertain. What does a bored cat look like? Cats sleep a great deal, it is true, but does this indicate boredom? Obviously not. If a human slept that much, we would think it was due to depression, but cats seem to need the prolonged rest in order to stay so alert when they are awake. However, an animal can be alert and still bored. Caged tigers and cheetahs, pacing back and forth, are clearly bored. How could they not be? They are not frightened or hungry, so their pacing can only indicate a kind of restlessness that is a basic component of boredom. What they lack is mental and physical stimulation. Their curiosity needs to be aroused, but it is difficult to provide excitement for an animal in a cage when it was meant to live in the real world. Animal boredom is no longer considered a joke; it is a concept of major concern to animal behaviorists, especially animal psychologists who work in zoos. Serious scholars have devoted their lives to working out ways to enhance the emotional lives of captive animals. Enriching their environment has become an entire scholarly field, with many publications. No scientist today would question the existence of boredom in animals.

No simulacrum can ever replace the adventure of living in a forest or jungle. In their natural world, jungle cats must make a living—that is, hunt for food. Learning what to eat, how to find it, and how to acquire it is very time-consuming. It also takes a great deal of energy, intelligence, and planning. They must seek out a mate to carry their genes into the next generation, the purpose of all life, to which cats are no exception. They must avoid enemies, which means they must know them and their habits. Careful research is life preserving. In order to avoid becoming prey themselves, they must stay healthy and keep themselves fit. They must learn about their territory, what it contains, where there is shelter from rain and predators. If they are females, they must devote considerable effort to raising a family and teaching the young to become self-sufficient at an early age. Life is not easy for a wild cat, but it could never be called boring.

What we have offered the domestic cat as a trade-off for this life is our presence and the compensation that this can sometimes bring. Playing with our cats, awakening their curiosity, giving them reasons to explore, is all the more important to compensate for their loss of total freedom. It may well be that the curiosity we see in our domestic cats is a replacement, a kind of *faute de mieux*, of the excitement of life in the wild. When four of the cats and I go on our evening walk, are they simulating their ancestral days in the wild? Everything they see on the beach on this walk draws their attention and awakens their curiosity. The same objects that they linger over—shells, driftwood, lava rocks—if brought into the house, would not focus their attention the way they do when encountered on our walk. They have, of course, never known life in the wild, but the way they move on open ground along the beach suggests to everyone who sees them wild animals rather than just playful domestic cats. Maybe there is something intoxicating about the sea air, the smells from far-off places carried on the breezes, the volcanic islands not far off shore, that awaken some atavistic memory. Or am I the one engaging in fantasy?

There is more to curiosity, though, than the need for adventure. Going to the top of our hill is not really an adventure, since it is a routine the cats know well. The outcome is never in doubt. Perhaps, though, I am speaking for myself. With every crack of a branch, every birdcall, every dog bark, they stop, crouch low, turn around quickly, and investigate. They do not seem frightened, but they are certainly not indifferent and casual. For them, this walk is serious business. Are they simply imitating the real thing, acting *as if* they were in a real jungle on their own? It is difficult to know.

When there is a strange noise in the rain forest, one they have not heard before, or when we encounter somebody walking down the path as we walk up, the cats will often run up to me, for protection, it would seem. They are curious, but they are cautious. I think cats are more curious than they are adventurous; they are, after all, ever vigilant. So far not a single one of the five cats has gone what the Australians call "walkabout," where they disappear into the bush in search of we know not what. Many friends, though, tell me how their cat simply took off one day. They assumed that the cat was lost forever, but several weeks later she would come wandering back, curl up in her usual spot, and give no indication that she had ever been gone, let alone breathe a word about where she had been all that time.

Miki, though, has begun to sleep around. I feared it might happen. Our neighbor Helen, an artist in whose studio Miki likes to hang out, told me that he appeared in her bedroom a few days ago and spent the night with her. Other neighbors report the same. He claims droit du seigneur in all ten houses on the beach and moves among them, showing no particular loyalty to any one place, even ours. It is a kind of co-housing for cats, very desirable from his point of view and perhaps from ours, too, after we get over the initial hurt feelings. Should I think of Miki as being disloyal, or is he just the ultimate free soul?

Today Ilan and I awoke early, leaving Leila asleep with Manu, and wished to go on a walk with the cats. We called, "Walk, walk, walk," which brings them to the door in record time, and we all set out for the beach. The tide was low and we decided to walk to the next beach, around the slippery rocks along the shore. It is a long walk, and I was not sure the cats would accompany us the whole way. They did, and they were clearly delighted that we were going farther than we had gone before and to someplace new. They pranced, they gamboled, and they leapt from rock to rock. The tide was coming in, and the tidal rock pools especially fascinated Megala, who dipped his paw into them to see what would come up. We walked for an hour, interrupted only by a man and his daughter, who shouted to us as we passed that he had never before seen a pride of cats walking on a beach, the second person to tell me that in less than a month. I felt again like a proud dad. Ilan turned to me and said: "Poppa, in my next life, I want to be a cat." He has the right ambitions, this boy. I suspect at that moment he was thinking of their utter joy in being cats. He, too, is bursting with curiosity and would never voluntarily miss a chance to exercise it.

Do the cats think they are on a hunting expedition, rather like a pride of female lions? Is curiosity, like playfulness in cats, connected with the hunting instinct? When Minna sees a bird, she begins that strange chattering sound we all know so well in our cats. It involves a rapid clicking of the teeth, accompanied by a kind of whine, an overflow of excitement rather than, as some people seem to think, a call begging the birds to allow themselves to be caught. Three of the cats do it when they see any insect on the windowpane. It is not that the cats are merely curious about the insect; surely an instinct has been stirred. In *A Cat Is Watching*, Roger Caras suggests that the chattering action is quite possibly an imitation of the cat's killing bite. The cat, he says, is anticipating getting whatever is in view by the back of the neck and severing its spinal cord. It is true that cats never use this sound when observing humans, no matter what they might think of any particular individual. We never seem to awaken the hunting instinct in a domestic cat. (But when Leila, Ilan, and I visited an animal sanctuary in Northern California, we had a different experience with a large lion who had only recently been rescued. As Ilan, then three, walked by the cage, the lion crouched and followed Ilan's every movement with his eyes. He looked as if he were preparing to pounce. There was no question that Ilan was prey in his mind. It was sobering.)

In cats, curiosity can be more powerful than a hunting instinct. Z. Y. Kuo, in a classic article entitled "The Genesis of the Cat's Response to the Rat" published in 1930, proved that a cat must be taught to hunt. Kittens raised with a mother who killed rats in their presence killed at their first opportunity; kittens raised alone seldom did, whereas those raised in a cage with a rat never did. Familiarity, in this case, breeds affection, or at least tolerance. Surely, though, one reason this works for cats and not for other animals is that cats are more curious, especially as kittens, about other animals. Before their hunting instinct kicks in, their curiosity is awakened—and it can override just about any other emotion, including fear, anger, and jealousy. For example, when Miki was a kitten he cornered a large weta, a strange prehistoric-looking insect a bit like a cross between a grasshopper and a scorpion, unique to New Zealand. He was clearly frightened, but overwhelming curiosity drove him to return over and over.

If cats are curious, they also have some curious behavior. Hardly any cat is without an eccentric side. There are certain things cats do that seem tied to their sense of curiosity while remaining apart from it. Eccentricity is very common in cats, much more so than in dogs. We label the behavior as eccentric because we do not understand it. Perhaps, though, what we see as eccentricity is simply a different emotion, one for which we have no name. I have a deeply held belief that cats have access to emotions that we cannot recognize. How would I know there are such emotions if I cannot recognize them? Three and a half centuries ago, La Rochefoucauld said in his *Maximes*, "We do not admire what we cannot understand." Cats prove that this is simply not true. Eccentric cats, cats who do things that puzzle us, fascinate everyone. Such antics add to the mythology of cats as mysterious creatures, when all they may show is that we fail to understand what emotions are driving them and therefore what emotional satisfaction they derive from their strange deeds.

There is a hotel cat at the historic American Colony Hotel in east Jerusalem, where T. E. Lawrence, Graham Greene, Gertrude Bell, Peter Ustinov, Leon Uris, and other colorful characters have stayed over the years, sometimes for extended periods. Sophia is a gray former stray who has been taken care of by the last two Swiss managers of the hotel. She is usually to be found waiting patiently for the elevator on the ground floor of the old stone annex building. When the elevator door opens, Sophia walks in quietly and sits calmly, waiting until the third floor, never getting off at other floors. True, she could easily take the stairs, perhaps in less time, but why waste energy needlessly? When the elevator reaches the third floor, Sophia stands by the door so she can be the first one out. When the door opens, she walks out and proceeds to visit selected long-term guests, with a seeming preference for writers. Sophia climbs onto the laps of her preferred guests and allows herself to be stroked. When she eventually becomes bored with this, she gets off the bed and scratches the door to be let out. Then she goes down the hallway and waits by the elevator until the door eventually opens, however long the wait might be. Then she descends, gets out on the ground floor, and waits to be let out the front door for outdoor adventures. What Sophia does outdoors in east Jerusalem (Al Quds) remains a topic of speculation among guests, but the manager of the hotel, who told me this story, hopes and rather suspects that it contributes to the peace process.

Now, one could argue that Sophia is merely curious and that all curiosity in cats seems to partake of a certain ritual element. If it was pure curiosity that drove her, why the obligatory actions? Why the third floor? Is this merely habit? Has she something in mind? She cannot count, I would imagine, yet must be guided by a feeling. What is this feeling she gets when she maintains her habits, day after day, and is it that feeling that drives her behavior? If so, it is not like a drug, for it is not something that enters her body. Rather it is something that enters her mind. If a human did this, we would say they were crazy, driven by an inner demon or an obsessional neurosis to repeat something because of the symbolic significance. The number three would not be without meaning. To understand such a person, you would need to know the person's entire history, something that is almost never possible. Humans are labeled, a foolish and often dangerous procedure (it can lead to hospitalization). Cats, however, get the benefit of our doubt.

We do not consider Sophia neurotic, obsessive, or a fool. We recognize that we simply do not have access to her secret inner life, but we cannot deny that there *is* such an inner life and that she is deriving some kind of emotional gratification from her daily ritual. Her feelings are not eccentric; they are merely unknown and not even necessarily unknowable. To know them would require keen observation and an open mind. If we are convinced in advance that her behavior is an empty gesture, with no significance whatever, then we are unlikely to find meaning in it. If cats cannot be humble about what they do not know, at least we can.

A while back, a helicopter landed on our front lawn to deliver the ice plants to our roof. I took the precaution of locking all five cats into the guestroom, where they had a superb view of the entire goings-on. As the many people began arriving to prepare for the landing, all of the cats had their heads locked against the window, watching everything with enormous interest. They did not look scared; it was just that their imaginations were working overtime. Saki (Hector Hugh Munro) wrote that the cat "still displays the self-reliant watchfulness which man has never taught it to lay aside." However, when the helicopter actually arrived, and I stood outside their window and pointed up, they all looked to where my finger was pointing and saw the giant bird poised in the air. (From this I gather that their eyesight must approximate ours for long-distance viewing). The look on all four faces was one of pure amazement: "That is the biggest bird we have ever seen" was the unmistakable message from their eyes. Only Moko looked scared; you could read the panic in his eyes, and soon he darted from the window to the safety of one of Ilan's play tents that littered the floor. I dared not intervene. It is very difficult to detain a cat who is convinced there is danger.

Fear can extinguish curiosity just as habituation can. Confidence in our maternal care (if cats see us as mothers), however, seems to be a reliable solvent for feline infantile anxiety. Do they gauge the seriousness of a situation by observing the degree of our concern? If I seemed worried, would they worry as well, or do they form independent judgments of matters that interest them? The cats seem to glance up at my face when something of interest first comes into their field of vision, which of course could tell them volumes. They know that most things do not concern me unduly. On the other hand, we just saw how when I am worried about their safety, as when they are walking on the ledge of the high roof, they remain unconcerned. Perhaps they figure that some things they know and others I know.

I have mentioned my surprise that the cats were not more curious about the arrival of baby Manu. This was difficult for me to interpret. I expected great curiosity and got none. I also assumed that as the cats became accustomed to him, their curiosity would diminish. I was wrong on all accounts. As they have become used to seeing Manu in the house, their curiosity is beginning to surface. When we leave him lying on his activity mat, one of the cats is now bound to come and lie down next to him. This morning, Minna Girl reached out her paw to touch him, ever so gently. Watching this reminds me that we so often impose our own expectations onto cats, and when they do not conform, we decide that something is absent, where it just may be far subtler than we are used to grasping. The cats did not *show* curiosity about baby Manu, but now I realize this does not necessarily mean it did not exist. It may have been below the threshold of *our* awareness. We might argue that they are now developing a curiosity they did not previously have. It is equally possible, though, that we are too stringent in requiring that all emotions to be recognized in cats must conform to our notions of how an emotion must manifest itself and writing it off when it does not. I am making a plea here for more subtlety. It may be that cats are simply subtler than we are, more attuned to the less obvious.

There is a well-known cartoon by B. Kliban showing a cat sitting in a corner with a thought bubble over its head, and what it's thinking is an image of the corner. This is supposed to show how literal minded cats really are. As a cartoon this may be funny, but it tells us nothing real about cats. They no more think about corners than we do. Almost everybody who lives with a cat, however, recognizes this image: Our cat suddenly stops and stares intently at a spot where we see nothing. The eyes widen, and sometimes the cat appears agitated, spooked, and will often abruptly turn and run away. Some believe that the cat has seen a ghost. Others think the cat is hallucinating. I think the explanation is less supernatural, more super natural, but no less interesting. I think that what has happened is that the cat is gripped by a memory. Since cats see no more than we do (slightly less, in fact; I remember when we all ran down to the shore to watch the dolphins playing just a few hundred feet away and the cats looked but could obviously not see what we saw, even when I took them up on the roof), they are not seeing something outside our field of vision. If they are not remembering some past event, could it be a fantasy that is so real, it is as if it were happening before the cat's very eyes? A kind of hallucination? The cat is reliving it, but it appears that he or she is actually reseeing it. Hence the reaction of the eye. The widening of the eye is the same reaction the cat shows to something that awakens its curiosity at other times. I think cats are so curious that they are even curious about events they imagine (or remember) in their own mind. They do not seem to make the same distinction we do between real and unreal (though most of the time they do—cats are ultimate realists and not the supernatural creatures in the imagination of people like Poe) and paradoxically may have, as a consequence, a better grip on reality than do humans. There is no evidence cats hear voices. When my cats strain to listen to something, it is not internal music they are listening for, but sounds beyond my range of hearing. A mentally ill cat is a great rarity. Even more rare is a sound cat with no curiosity.

I began this chapter by saying that curiosity is not exactly an emotion but is not exactly not an emotion, either. Nonetheless, curiosity is closely allied with feelings. I have the sense that cats are curious because they want to know how something feels. We use a wheelbarrow to carry things up and down the hill. First Miki, then Moko recently began hitching a ride as I carry things up. They were curious, I thought, as to how it would feel to be carried in the wheelbarrow. Now it is common for them to catch a ride with me. They stand in the front, balanced a little precariously, their faces uplifted in the gentle summer breeze, sniffing the air, ears big and eyes alert. Above all, they seem to be feeling the pleasure that comes from curiosity fulfilled. It is a heady mixture.

NINE

Playfulness

Megalamandira, left, *and Miki*

I would argue, after observing cats over the course of my lifetime, that playfulness is one of their key emotional states. Play gives rise to emotions, yet we do not have names for the feelings that come up in us as we play, perhaps because play occupies us so little as adults. For children, though, right until adolescence, it is their major occupation. The same is true for cats.

Cat play has been less studied than dog play, which is too bad, because it is fascinating and by no means as simple to comprehend as might appear at first glance. Since four of my five cats spend most of their time playing, and I am home all day observing them, I have had an unusual opportunity to study cat play.

It makes sense that much of the body language exhibited by cats to express friendliness is the same as that used for play. Whenever I approach Miki, the moment he sees me his tail goes straight up in the air and he comes quickly toward me with it held high and straight. At first I did not recognize this for what it is—a sign of friendship mixed with confidence, but now that I have seen it, I see it everywhere. Happy cats lift their tails up, and judging by how they modify their behavior, immediately responding to the elevated mood as if happiness were infectious, we know that other cats understand this. Of course, if other cats could not understand the gesture, it would never have evolved. It is as clear in its meaning to a cat as a handshake is to a human. As many cat observers have noted, and I have seen it as well, cats who bend over the end of the tail in a half question mark are signaling friendly intent, but without the confidence. Similarly, an arched back is a reliable sign of play, but when accompanied by other gestures, such as hissing or a hostile stare, it indicates aggression. Kittens often open their mouths when soliciting play. It is possible, too, that certain other gestures—for instance, the half crouch and the pounce—signal the end of the play session. One researcher says that the horizontal leap, where a cat suddenly moves sideways very quickly, is a way of saying, "Play is over." These signals are meant for other cats, not for us, and we often miss their significance.

Still, it seems to me that not all cat signals have the singular clarity of the ones made by dogs. Dogs have the play bow, which is a simple way of saying, "Let's play." If at any moment during play the bite is too hard, or the body slamming too strong, dogs will revert to the play bow. My friend Marc Bekoff, a professor of biology at the University of Colorado at Denver, tells me this is a dog's way of apologizing, of saying, "I'm sorry—it was a mistake, I was only playing." Cats have the equivalent of a play bow: Minnalouche constantly urges me to play by hurling her little body on the ground in front of me, flopping down, stretching, and reaching out with her paws. It is very effective, for how can I resist such a request? However, cats seem to lack the mechanism that allows them to apologize—they do not roll and stretch when they make a mistake. Therefore cat play can get them into more trouble than does dog play with a dog.

As we have seen earlier, cat play-fighting frequently turns rough and then proceeds to real fighting. Cats begin by grabbing each other by the neck and biting, but one bites too hard, and the other is hurt, then turns around and bites back harder, and soon there is a fight. The reason they seem to lack the canine ability to apologize is probably that their wild ancestors, being largely solitary, have not needed this ability. To whom would they have apologized? Cats, lacking a play bow equivalent, stalk off, looking ruffled and miffed, as if vowing never to play with the offending creature again. Of course they do, and sometimes just minutes later, so they must have some more subtle mechanism of recognizing that the fight was not intentional after all. I have watched Miki and Moko, who are close friends, begin a harmless play fight, only to see it escalate and within seconds career out of control, with what sounds to me like horrible screaming and wide-eyed terror. A minute later they are licking each other. The licking does not feel like an apology, more like forgetfulness, since cats live in and for the moment. Moko will begin to play with Minna, but the play goes too far, and there is a fight, though not a nasty one. It seems to me they will avoid each other for the rest of the day, but I am wrong. Moments later they are lying close together on the couch and taking turns grooming one another, as everyone sees repeatedly with cats.

The rules of the game are not nearly as clear to us as is the case with dog rules, though obviously cats themselves seem to understand them fine. Moko is sitting on my lap on the couch while Yossie grooms him. Yossie licks Moko's head carefully and tenderly; suddenly he bites his neck. From where I sit, it looks like a hard bite, and it seems to be getting harder. Moko agrees: he jumps up, looks startled, and bites back. "Uh-oh," I think, "the grooming session is over and a fight has started in earnest." However, I am wrong, because just as suddenly as the "fight" started, it ends, and now Moko is grooming Yossie. Have I missed something? Undoubtedly.

Moko seems to know when I make an honest mistake and to accept the fault good-naturedly. I mentioned earlier that when we run down the path together, he has a habit of getting in between my feet. This is his way of playing, and he does not expect me to make mistakes. But sometimes I nearly trip over him and kick him accidentally. He jumps to the side with a light step but looks up at me to let me know for certain, though I would never make the experiment, that were I to kick him for real, he would not show tolerance. The trust I have slowly built up with him over time could be lost in a single gesture, I fear.

Minna's understanding of an apology seems to go further than Moko's. When Leila stepped on her foot by mistake, she went screeching from the room. When Leila told her how sorry she was, Minna came running back in, rubbed against Leila's leg, and purred loudly. What else could it mean, but "I accept your apology, I know you didn't do it on purpose."

Why do adult cats play less than kittens? We cannot say that an adult cat has other, more serious preoccupations, such as hunting or keeping safe, since we provide food and safety for cats. I have never seen one of the younger cats refuse to play when solicited by one of the other cats. This agrees with research showing that it is very rare for a cat to respond to playful gestures with any kind of hostility. But that same research shows that mother cats can become annoyed if their kittens are overly playful and will show their annoyance by hitting them with their paws or growling at them. If none of that works, they will simply walk away.

Yossie rarely plays anymore with the younger cats. If they persist, he seems to agree to play, but in fact he plays so rough that they run away, offended. He is clearly hoping they will not continue to bother him. He is too old to play. What does that mean? Could it be that just play per se no longer has much significance for Yossie, that he has nothing more to learn from it? This would make sense in terms of the evolutionary significance of play in the life of cats. Meredith West, professor of psychology and biology at Indiana University, hypothesizes, and I would agree, that play might have evolved because kittens who play together remain accessible to the returning mother but safe during the time that she cannot attend to them. Feral mother cats usually leave their kittens when they are asleep or playing, so that the kittens do not attempt to follow the mother. Contrary to stereotype—such as pictures of mother cats wandering around holding a kitten by the scruff of the neck—mother cats never travel with their kittens; this would expose all of them to danger. If West's theory is true, then the generally held view that kitten play is invariably related to later hunting behavior loses some of its cogency.

No doubt the only reason an adult domestic cat plays more than a feral adult cat is that we keep our cats in a more kitten-like environment. Why do we find it particularly charming when an older cat still plays like a kitten? Probably for the same reason that we enjoy an adult human who plays like a child. Play is an indicator of happiness. A depressed person and a sad cat do not play—though if they played together, it might benefit both. When I see Yossie begin to play, especially with the other cats, I assume he is feeling less morose about life. Play releases endorphins in humans and cats, and we all need that. Play and laughter fertilize the brain, the neuroscientist Jaak Panksepp reminds us. Cats may not laugh in identical ways to us, but their play causes us to smile and laugh and is often seen as the equivalent of pure laughter.

I love to watch Miki trying to frighten off one of the neighbor dogs: he prances up to him, then suddenly turns sideways to show him his biggest side, then continues dancing forward. It is not that the giant dog is frightened, but perhaps he is slightly unnerved or puzzled by this odd-looking animal coming at him by moving sideways like a spider. He backs off, and the cat relaxes. I see the cats make this same odd movement when they are play-fighting: coming at one another sideways, their hair horripilating (standing on end, also called piloerection) to make themselves look large and ferocious. Which they do look. It is definitely play, but it is easy to mistake it for the real thing, for two cats will bite each other and roll around the floor and use their back feet to tear at the gut of the other cat. It is very realistic. Nevertheless, it is not real, and they know it. How do they?

Moko has a game he plays with me: We walk together down the beach to a very old (possibly six hundred years) pohutukawa tree (called the New Zealand Christmas tree because it breaks into red blossoms in December), with enormous branches that reach right down to the beach over the water. Moko leaps into the tree and then waits for me to tap a branch with my finger. He then leaps onto that branch and watches my face. When I tap another one, he leaps there. He is purring the entire time. He likes taking orders—as a game. When he is finished, he is finished. I can insist on playing some more as much as I like, he just walks away. You cannot argue with a cat. He is not obeying me; he is playing with me. Perhaps he knows that this is a game he could not play with another cat, only with me, and that gives him pleasure. It gives me pleasure, and I wonder if it might not be pleasure of the same order for both of us: we have, each of us, succeeded in engaging in play a member of another species. It is not a game he would play with cats or I with another human, but it is a game we play with each other. By breaking the species barrier and playing with an alien, this play may simply be another way for cats to express their love for us, their feelings of affection and friendliness. This morning, when the game was over, he rushed up to me and pushed his head against my arm, purring wildly. Could anybody who observed this doubt that he was not only glad to be playing, but feeling affection for me?

Moko and I also play a game called *un elefante*, after a Spanish children's song in which the children walk along a ledge or stone wall and a person sings about an elephant who balanced himself on a spiderweb. Ilan loves to play this every time he sees a fence he can walk along. Moko has watched our game and was apparently determined to put his observations to good use. Whenever we come to a fence on our walks, he leaps up and walks along, looking at me with something akin to triumph. Of course, I sing the Spanish words for him. (I think he prefers it when I do not sing; after all, he likes Bach, so my singing must grate on his ears.) He does not like it when I try to steady him or help him in any way, not that he ever needs it in any case.

In the game I mentioned earlier—where the cats leap in front of me as we meander down the path and then suddenly flop down full length stretched out in front of me so that I nearly trip over them—it is clear they must be conscious of what they are doing, for they never do it with one another, only with me, as if they knew it would make no sense to do this with another cat. The game they have come to play increasingly with one another is for one cat to come rushing out of the bush and leap over the other cat's back. They have not tried that one with me, so far.

Miki did something similar with Ilan and his bicycle. He stretched himself out on the narrow path in front of Ilan, thereby taking up most of the room for maneuvering. I assumed, and so obviously did Ilan, that as the bicycle approached, Miki would jump up and leap nimbly out of the way, the way dogs do when lying in the path of a car. He did no such thing. He simply lay there, watching with anything but apprehension as the bicycle careened toward him. Ilan was barely able to swerve aside onto the grass at the last moment to avoid hitting the perfectly nonchalant Miki. Was this some obscure form of challenge or just a joke he was playing on himself? I do not know, but he was not about to budge for anything and clearly was enjoying himself at the same time. It was dangerous for him; the bike is large and would have hurt him if it had run over him. While he may have known, it was as if his sense of play were more powerful, at that moment, than any thought of safety, a most unusual state of affairs for a cat.

Miki is perhaps just more playful than the other cats. The other day he got under the small trampoline, which the kids were using next to the large one. They would jump from the big one to the small one, and then Miki would stick out his paws and try to grab the children's legs. He looked up to see them bouncing right over his head. The kids were trying to scare him, but he was doing the same to them. He seemed to enjoy himself enormously.

As I have mentioned, the cats understand the word *walk*. When I say it, their ears prick up, and four of them (Yossie never rushes anywhere) rush to the door. I have been going with them every evening to the giant old pohutukawa tree. Today it was raining hard, but all the cats were looking at me expectantly: "What's the problem? It's five o'clock." So I went, and though the rain poured down, they were right there, as jubilant as ever. Moko immediately rolled in the sand, then dug a hole, raining or not, then raced after Miki and somersaulted over him to land with a thud in front. No wonder they look forward to a "walk" and know the word. Cats seem to know only the words for pleasurable activities. Mine know "treat," "food," beach," "swim," "tree," and "walk." And their own names. (However, they do not seem to know one another's names; they never look to the cat I am calling.) They will never learn "no" or other unimaginative negative words, or perhaps it is just that they will not acknowledge that they know the meaning. Cats are masters at ignoring you and your vocabulary when they feel like it.

Since the cats take such enormous pleasure from our daily walks on the beach, playing with one another and with me, the question arises as to why they do not do it on their own. Why do they wait for me; why don't they decide to take a walk together, without me, for the sheer fun of it? I must add something. Since cats, especially when it comes to fun, are nonhierarchical by nature, they can hardly mistake me for their leader. However, I do provide some kind of incentive they seem to need. They would probably have just as much fun on their own, but they do not go. They need me. By nature solitary, my cats are nonetheless part of a group. It must be a new kind of pleasure for them, a result not of evolution but of domestication.

Yesterday I tried an experiment. As I was feeding the cats, just as they began to eat, I announced that I was "going up the hill" for a walk, then set off. Torn, three of them, Miki, Moko, and Megala, bolted down as many mouthfuls as they could, then set off with me! So they prefer being on an adventure with me to eating. Pretty remarkable, but what does it mean, since they could perfectly well go for a walk whenever they want on their own? It is not that I am top cat, but something about the combination of walking and me wildly appeals to them. They never pass up the opportunity. As they become increasingly free to do what they want, they appear to regard me as having been reborn as a cat.

The cats are so used to waiting for me about halfway down the path, hiding in the rain forest, that when they hear my footsteps, they begin to call. When I answer, they rush out of the forest and race toward me, behaving in every respect like dogs. Even when they are at home, when they hear Leila, me, Manu, and Ilan approaching in the evening, they rush out the cat door and meet us halfway, running up to us and rolling at our feet. They might have been watching the neighbors' dogs, who also do the same. I may be just lucky, or they could be conforming to my own expectations. I mean, knowing that I like this so much, have they provided it? Is that at all possible?

We had ten kids over after school a few months ago. Ten wild five-year-olds. Four of the cats were downstairs when the children raced in the door: three of them took off immediately, up the stairs and into secure hiding spots. Not Miki. He stayed where he was, cocked his head in that way he does when he is curious, and walked up to the first kid, who scooped him up right away. Then twenty little hands began probing him, tickling him, lifting his lips, feeling his gums, and tapping his teeth. I went to save him, when I heard that little motor going: Miki was purring and enjoying every minute of it. For the next three hours, the kids continued to play with him, carrying him upside down, legs dangling, and through it all, you could hear that loud purr of contentment. He not only did not mind, he positively loved it. To my surprise, after about an hour, the other four (even Yossie!) emerged and began to take an active role in the children's play. I fully expected them to have stayed in hiding until the children were gone, but the play clearly appealed to them.

This was not a unique occurrence, and now our five cats seem to be on their way to becoming hypersocialized. On August 28, fifteen little boys and girls invaded our house for Ilan's fifth birthday party. They ran about screaming, yelling, and throwing things. To my surprise, instead of disappearing immediately, each of the five cats came out and stood in the middle of the group. Moko was especially curious, looking from one child to the other to see what would happen next. Miki, of course, allowed himself to be carried in every undignified posture, his feet hanging indecorously, but good-natured as usual. Yossie, with his enormous green eyes, took it all in as well, and Minnalouche lay posed elegantly in the middle of the porch. Even Megalamandira, whose first exposure this was to such a gang, was brave and pleasant. When a child would position himself in front of one of the cats and jump up and down shouting, even I would have turned and fled, but the cats stood their ground, showing no distress, merely placid inquisitiveness. They have been exposed, on nearly a daily basis, to such a wide variety of people and events, such as the long beach walks, that they now seem to crave the novelty and the excitement. They have become more than socialized; they are becoming thrill junkies! They seem to recognize that what the children are doing is play, and they love play.

Now that all the cats are happy to play with one another, they still play with me. They like it when I come down from my great height and play at their level. They are not always play-stalking. Sometimes they are playing according to rules I devise now, seeing if they can understand what I intend. They do. Furthermore, they enjoy it. There is, it would seem, a kind of play that is divorced from any purpose or use in the real world. It is the sheer joy of play. Yossie, being older and more dignified, plays less than the other four, who are all adolescents now.

Are cats playing in a special way when they play with us? It seems they make concessions for humans they would not make for another cat. They assume some sort of handicap on the grounds that we are—well, play-challenged. What, though, are they thinking when they play with another species, one to whom they are beholden for food, shelter, and comfort? A few days ago, I raced down the hill to the house, and Moko and Miki stayed behind, in the rain forest, without my noticing. Later, as I was checking my e-mail, I missed them. About an hour had gone by, and I walked up the hill calling them. When they heard my voice they emerged from the trees and ran toward me, calling loudly, their tails held high, and rubbed and rubbed against my legs, as if to say, "We thought you were gone!" Were they just pretending (as part of the game) that I had left them? It is not easy to tell. A horrible thought: Is what I regard as a game serious for them—our walks, for example? Could they be following me because they are afraid I am abandoning them? Sometimes the cries they make when I leave perplex me. They often sound so plaintive. Am I missing something or are we all missing something? Dogs, after all, have never been able to understand why we would ever leave them. I assumed this was because they were, basically, wolves, and wolves don't leave one another. Cats, I thought, would want to be alone. Have they evolved into some other creature, resembling a wild cat but actually quite different in their emotional needs?

Undoubtedly there are clues to what a cat feels in the different sounds they make. Humans are poor judges of these sounds. When I return from an errand and call the cats out of the rain forest where they have been waiting for me, they call back. What do I hear? A loud, plaintive, almost wild sound. The part of the sound that means "Here we are!" is unmistakable. But it could be that there is a complaint mixed in. They might be asking, "Where have you been? How could you leave us like that, alone in the forest?" It is difficult to interpret the sound. They could be playing, teasing, or chastising me. They are certainly not scared when they emerge onto the path: they come running and then roll at my feet, purring loudly. Perhaps this is a purr of relief that their expectations have been met and, against all odds, I have returned. I suspect the different sounds they make, both the calling and even the purring, would tell us, if only we could understand them.

You can never judge how happy a dog is, because with a dog it is always exuberance in extremis. A five-minute absence is greeted very much like a two-day absence. We can agree that cats are more measured than dogs in the level of joy they express in play, but each cat seems to have his or her own level. At least three of my cats, Moko, Miki, and Megala, behave more like dogs than cats in that they are prepared to follow me wherever I go, as often and as long as I wish. They remind me of Ilan and, like macho little five-year-olds, are always ready for adventure. Minna gives me the more feminine pleasure of feeling adored. Yossie makes me feel like an intruder into the world of cats. He retains his mystery to the end.

At first when they would follow us up the hill, they returned on their own immediately. Now, if it is late—say, if we have gone to a movie in the nearby town—they wait for us. When they first did this, I thought: "What can they possibly do in the dark, waiting for three hours? How boring for them." As we were getting into the car, however, I saw Moko and Miki leap into a tree in the rain forest that borders the path. Suddenly I saw the wait from their perspective: three hours of play in a rain forest at night, sheer bliss!

Last night I walked Ilan's (Manu still comes with us everywhere) baby-sitters (he is so popular that three teenage girls like to stay with him) up our hill, at eleven P.M. It was a quiet, moonlit night, with nobody around, and when we left the house, all but one of the cats walked out with us. I expected only Miki and Moko to go all the way, but to my surprise, four of the cats joined us (Minnalouche was asleep). They were in high spirits: they kept playing leapfrog, got more and more excited (they seemed intoxicated by the moonlight), and raced up trees along the path and dashed down again, ran through my legs, and stretched their little bodies out along the path. I chased them; they chased me. It was clear that I was amusing them. I have rarely seen them so exhilarated. Clearly they reach the height of playfulness when they feel safe in a peaceful setting. Maybe they were reminded of some earlier pleasurable experience and were invigorated the way I was by merely seeing them.

• • •

That kittens need to be exposed to play is obvious if one watches a cat who did not have the opportunity to play as a kitten. Studies have shown that such a cat, as an adult, does not respond to the appropriate play signals. Solitary kittens, if they have been exposed to play, will often play with imaginary playmates, making the same precise gestures they do with real friends, and sometimes this persists into adult life. When we see an adult cat suddenly leap in the air, we think he is hallucinating. More likely he is playing with an imaginary friend. Some vets maintain, and I agree, that a kitten should not be removed from the mother at six weeks, which is very common, because the peak of playful interactions comes at eleven weeks, and cats who stay with their mother and their siblings until then will remain more tolerant and sociable in adult life. Twelve weeks, then, would be the ideal age to take a cat from her mother, who will probably express her gratitude. It is very hard to undo the result of those early experiences.

I wonder what the cats do when I am not there. Do all five spend their time in a state of suspended animation, waiting passively for me to return? I have looked through the window and seen them all sleeping peacefully enough on the couch as I walk up the path to the house. As soon as they hear me, however, they leap off the couch, and the fun begins. They seem to have been waiting for me. I animate them in some peculiar way and for reasons I cannot quite fathom. I do not understand what I provide that they do not have with one another. Perhaps there is a special pleasure in playing with an alien species. It could intrigue them as much as it does me. Along these lines, I find that two of them in particular, Moko and Miki, are getting more and more interested in the three neighbor dogs, a different alien species, who spend most of the day lying in the sun on our front deck. At first, all the cats avoided them as much as possible. Now, however, they are inching up to them, clearly interested. In what might that interest consist? I think—and only time will tell if I am correct—that they want to play with them. In fact, I was sitting out on the beach with the cats a few days ago when a neighbor picked up her videocamera and began recording an odd scene: Miki was inching her way up to Roxie, a small ginger dog who lives next door. Roxie was staring out at the ocean (her favorite sport is to race into the water chasing the shag birds—she will swim for hundreds of yards out to sea in the vain hope of catching one). Miki lifted a paw and then swatted Roxie's tail. Roxie did not move. Miki did it again, and Roxie, annoyed, stalked off. I think Miki was initiating play. However, if there is pleasure to be found in cross-species play, why has not more of it been reported from the wild, where few, if any, authenticated cases have ever been observed or recorded? Is it rare or simply unsought because not expected?

Unlike purring, play is not always social—that is, I see my cats play alone as happily as they will play with one another. They are not necessarily playing to an audience. However, I have noticed that solitary play (chasing shadows, for example, and perhaps I should include watching TV since many cats enjoy seeing small animals on screen, especially birds and mice; there are videos for homebound cats) is never as prolonged or, to my eyes, as intense as social play. Since cats are, by evolution, solitary creatures, the question that is almost impossible to answer is whether any adult wild cat engages in play. I am not aware of any recorded instance—whether video or hidden observer—of this. I suspect that a wild cat will sometimes pounce or chase an object for fun, but that it will be short-lived. Something that cats have gained by their association with us is an increase in the intensity of their play life.

In the evening, the five cats begin a crazy dance where they rush up the hill behind our house, madly climb thirty feet into a tall cabbage tree (that resembles a giant palm tree) that we have out back, dash down, then race up the hill and down, tumbling, jumping on top of one another, and showing such joie de vivre as I would not have believed cats capable of. As my friend Michael Morrissey, the New Zealand novelist, put it to me, the ideal feline is part canine!

In fact, Alexandre Dumas had a cat with the mysterious name of Mysouff, who would meet the famed author at the end of the day and walk home with him. Dumas explained: "That cat missed his vocation, he should have been born a dog." The Dumas family lived on the rue de l'Ouest. Each day, Mysouff walked Dumas to work as far as the rue de Vaugirard. On the writer's return, his canine cat would greet Dumas: "He would jump up on my knees as if he were a dog, then run off and return, then take the road home, returning a last time at a gallop."

Biologists have long been interested in the evolutionary significance of play, its purpose. Much has been written about how the play of cats resembles hunting technique: they lie in wait for one another, pounce, bite the neck, and push with their feet, all as if they were stalking, attacking, dispatching, and disemboweling their prey. If this is true, it is not surprising that the play sometimes rolls over into the real thing, or what looks like the real thing. Is it true, as Elizabeth Marshall Thomas claims in her book *The Tribe of Tiger*, that "the behavior of kittens at play is hunting behavior and nothing else"? Reluctant as I am to disagree with my favorite author, I think not. Today, the cats demonstrated a completely new trait: they invented their own toys. They cooperatively shredded some cardboard and made little balls out of it, which they then swatted back and forth. The swatting we could relate to hunting behavior, but the making of the little balls they clearly did solely for the enjoyment of it. Just yesterday, Megalamandira came prancing out of the closet with a toy mouse in her mouth. When Minnalouche went to investigate and join in the game, Megala, for the first time I have heard, growled a warning. This was his prey, and nobody was going to take it away. Minnalouche looked taken aback, as if Megala were forgetting that he was holding only a toy. In this case, Megala was clearly exhibiting hunting behavior and following the appropriate rules: No cat is allowed to take away prey that is in the mouth of another cat, no matter who is top cat or how the hierarchy is arranged. Moreover, cat hierarchy, as we have seen, is highly fluid.

Cruelty is so often attributed to the cat that we must consider it. Cats do play with mice before killing them. However, in order to label them as cruel, we would have to believe that they derive pleasure from watching the mice suffer. A sadist derives pleasure from someone else's pain, but a cat is not a sadist, for there is no evidence (anywhere in the animal world, by the way) that any animal is sadistic by this definition. The cat may be cruel by human standards (by cat standards we are far crueler), but not by his own. Picasso, who loved cats, nonetheless painted a cat showing sadistic pleasure in cruelty, *Cat Devouring a Bird*, saying, "The subject obsessed me—I don't know why." The animal holds a vainly fluttering bird in its fangs. However, Picasso painted it in 1939, during the Spanish Civil War, so it is not hard to think of the painting in allegorical fashion, with the cat representing fascism and the bird freedom. One of the reasons we dislike watching cats kill other animals is that there is symbolism (of cold detachment) involved, not to mention the actual event itself. However, we rarely consider our own shopping for chicken (a bird) in the supermarket to be a cruel activity, but somewhere along the line, it was. We are not so different from our cats. Or rather, our species is certainly far more evolved when it comes to killing and cruelty; no cat has ever set out to destroy every other cat from a different lineage. Genocidal fury has never consumed a cat, only a human.

If we are crueler than cats, we nonetheless seem to possess more humor. I tend not to think of humor as something that cats possess in abundance. Playfulness, yes, but humor seems somehow, well, a little bit undignified for a cat. More of a dog thing. Yet Yossie has a sense of humor. There are nights when the cats are simply too active for us to get any restful sleep, especially Ilan, so we close the bedroom door. We leave the balcony door open, since as far as we have been able to ascertain, there is no way onto it from anywhere but the bedroom. Yet in the middle of the night, suddenly Yossie materializes with a loud greeting that sounds to me like a chuckle. He asks to be let out through the bedroom door, which I do, and he gives me a triumphant look: "You'll never know!" An hour or two later he is back in the bedroom, just to make his point clear, and this time he stays to sleep on Ilan's bed. He is playing with us, and he seems to know he has stumped us. Nevertheless, I finally figured it out: He makes an impressive leap from the garden to the roof. The point remains that he seemed to know I did not know how he did it, and he enjoyed my puzzlement.

Miki was walking along the seashore when a huge seagull landed just a few feet away from him (deliberately teasing him?). Miki went into hunt mode: crouch, crawl, and sneaking up. The gull moved farther onto the mud flats; Miki followed, shaking his feet as he stalked the bird through pools of tidal water. There were young seagulls out there; you could distinguish them easily from the adults by their markings—they were gray, while adults are white and black. (This visible distinction between the fur of young animals and adult animals, very common in many species, is absent in cats; the markings don't change as the cat matures, which means, I think, that a very young cat is, for practical purposes, as mature as an adult.) More gulls arrived, perhaps to protect the young. Soon Miki was surrounded. He was playing, but the gulls were not: they whirled in the air and began dive-bombing him. This was too much for our little hunter, and he raced up the beach for the safety of our house. He ran inside, stopping only to pick up a large leaf and carry it in his mouth as if it were prey. He presented it to me. Does he have a sense of humor? Or was he saving face?

Play will expand exponentially in response to opportunity. The reason we see so little variation in cat play is that we place so many restrictions on its expression. Cats will, of course, play with a ball of string or chase their tails. Stampeding behavior, where the cat races from one end of the room and back again, is common in an apartment as well. But these forms of play are compensation play. Cats have wonderful imaginations, and if deprived of their natural environment, they will improvise well. If I could live my life over, I think I would like to become an architect specializing in devising imaginative playgrounds for children. They are, at present, so conventional: a seesaw, a swing, and a slide. They could be so much more. The same is true for cats. However, no playground, for children or for cats, can replace the natural world.

I get almost inexpressible pleasure from walking with my four and sometimes five cats on the beach. Today it was an unusually rainy, foggy, cold day here. The cats sat, all of them, on the sofa. It seemed they were waiting for something. As soon as I said, "Should we go for a walk?" four of the cats pricked up their ears, then jumped off the sofa and walked to the door. Out we went. Once we were on the beach, in spite of the drizzle, they were ecstatic, chasing one another, climbing the old pohutukawa tree on the beach (that has just burst out in red blossoms), leaping from branch to branch, racing up the side of the nasturtium-covered cliff, running right up to the huge waves (it is a very stormy sea out there today; you can't even see the islands just two miles from our house), with Megala, a leopard after all, dipping his feet into the waves and altogether acting as if this were the first and quintessential outing of their entire lives! I could not get enough of watching them, even though we do this every day. Why does it give me such a pleasure? Partly, I suppose, because it is so unexpected that four of them love to walk. Partly it is because they look so spectacular on the beach, these four dots of bright colors, all rolling around in the sand, jumping in the air, and leaping about like drunken frogs! They could go out any time, yet they wait for me. They look so disappointed (not an easy emotion for a cat to express—but they fix me with a blank expression that borders on contempt) when I say it is time to go back home. They drag behind, reluctant to stop for the day, but they follow me back anyway and then settle on the couch, once again closing in on themselves and looking mysterious and self-satisfied, the way cats seem to do. I am left to puzzle out why this whole scene, day in and day out, will never cease to enchant me, and apparently them.

How could we fail to see that just as some animals make us envy their remarkable physical capabilities, so do others make us wish for their emotional gifts. Cats have an emotional directness, a purity of purpose (or is it purposelessness?)—no compromise, especially with the honesty of their feelings.

Here I am, every moment of the cat says, love me as I am, or not at all.

With my cats I am learning the lesson of the sufficiency of the moment. No yesterday, no tomorrow, only the magic of today, of this single instant. No remorse, no regret, no yearning, just the play of now.

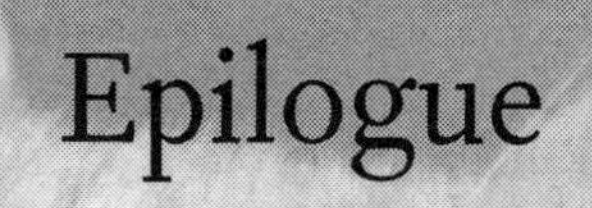

Epilogue

Miki

Yossie went to live with Jane and 120 other cats for three weeks, to see if there was something here that was making him unhappy. He returned today, March 19, 2002. He promptly bit me. Was he letting me know how he felt about being sent away, or was he back to his usual surliness? Ilan simply will not permit me to send him off again, so I will have to live with a bad-tempered cat. Perhaps I will learn to read him better.

I have lived with the cats for one year. I realize now how long that is in the life of a cat. They have grown from joyful kittens into serene adults. It is something of a shock to learn how quickly they have matured to become ever more mysterious beings, with a private core I cannot penetrate.

I can no longer consider Miki "our" cat. It is not that I wanted to possess him, but he clearly belonged here, in this house, with his four feline friends and us. Or so I thought. Evidently I was wrong. He is his own self now, an independent individual. He sleeps where he likes, hardly the same house twice in a row. Were someone to come calling for him, I would have to say he visits us but no longer lives here. Where, then? Everywhere and nowhere. He has left home, like any eighteen-year-old, for that is how old he is if years are measured in maturity. I can interest him (barely), but I cannot own him. Of course, I never did. What a lesson in humility for me!

The cats, as they mature, seem to view me as a friend, perhaps even a likable relative, but no more than that; I am no longer the center of their lives. What has happened? They have matured, true, but they also have become free. They have not physically become feral cats, but in some important sense they are now emotionally feral, especially in the sense of implied freedom. They come and go as they please, and this has given them a kind of emotional as well as physical independence. They feel as little or as much as they like, for us or for anyone else. It is not that they have replaced me—they have merely expanded their own lives.

Still, I cannot leave the house and begin walking along the beach or the path leading through the rain forest to the top of the hill without all the cats materializing magically to follow me, single-file. I do not have to call; in fact, I can be as quiet as I like in making my exit; *they know*. Their exuberance, their spirit, is never more evident than at these moments. They preen, they call, and they hold their tails high with pride as they march to the top, my faithful band. It never fails to move me, to enchant me. At the top they disperse as mysteriously as they grouped and continue their lives singly and as individuals. For those rare moments we are united in something I cannot entirely understand but whose value to me is beyond precious.

Elizabeth Marshall Thomas called her book about dogs *The Hidden Life of Dogs* because she wanted to know not so much about their emotional or inner lives as about what it was dogs did when they were free to roam independently. The lives my cats lead apart from me is the hidden life of cats. I catch glimpses of them on the beach, at a neighbor's, by the giant tree. They come round regularly to greet me, sometimes to sleep here, to eat always, and often just to rest in the shade under the lounge chairs on the deck facing the ocean in the summer heat.

Sometimes they will call me, as if inviting me to a special adventure, different and more enigmatic than our communal walks. But I cannot decipher their sounds and so must decline. They shrug and set off individually to I know not where, and I am left with the feeling that I have missed out on something essential, some clue to the mystery of these perfect beings who briefly and mysteriously grace my life.

Notes

INTRODUCTION

xvii *Cats "sought rather than bought"* According to the Humane Society of the United States, eight to twelve million cats and dogs come into shelters annually in the United States. Of these, four to six million are euthanized. Less than four percent of cats are reunited with owners and less than ten percent are adopted out. See www.psyeta.org.

CHAPTER TWO: LOVE

35 *"Cats" by Robley Wilson Jr. The Company of Cats: 20 Contemporary Stories of Family Cats,* ed. by Michael J. Rosen. (N.Y.: Gallahad Books, 1996), p. 28.

38 *sensitivity of a cat's paws, whiskers, etc.* The whiskers, known as the vibrissae, are found on many parts of the face, as well as on the front (but not back) legs. The bases of these hairs are well supplied with

nerves and are very sensitive to touch. A substantial nerve system transmits tactile information from the vibrissae to the brain. It is not a myth that cats use these ultrasensitive hairs to determine whether they can emerge from a small space they wish to enter.

50 *the cat was neurotic* Sue Hubbell: *Shrinking the Cat: Genetic Engineering Before We Knew About Genes* (Boston: Houghton Mifflin Co., 2001), p. 113.

55 *story of Scarlett the cat* Jane Martin & J. C. Suarez: *Scarlett Saves Her Family* (New York: Simon & Schuster, 1997).

CHAPTER THREE: CONTENTMENT

67 *The cat amid the ashes purr'd* John Wolcott (Peter Pindar), 1801 (quoted in vol. XII of *The Oxford English Dictionary*).

76 *"While the shadows grew long"* Elizabeth Marshall Thomas, *The Hidden Life of Dogs* (N.Y.: Houghton Mifflin, 1994).

CHAPTER FOUR: ATTACHMENT

86 *I had the pleasure of seeing firsthand* Bob Walker, *The Cat's House* (Kansas City: Andrews & McMeel, 1996).

88 *indoor cats vs. outdoor cats* Cleveland Amory, *The Best Cat Ever* (Boston: Little, Brown & Co., 1993), p. 223.

90 *Jeremy Angel* Jeremy Angel, *Cats' Kingdom: A Long and Loving Look into a Very Special Feline Community* (N.Y.: Warner Books, 1985).

CHAPTER FIVE: JEALOUSY

111 *a number of studies of feral cat colonies* Paul Leyhausen, "The Tame & the Wild," in *The Domestic Cat: The Biology of Its Behaviour*, first ed.,

by Dennis C. Turner and Patrick Bateson (Cambridge: Cambridge University Press, 1988), p. 62.

113 *"She never showed any signs"* Howard Loxton, *The Noble Cat: Aristocrat of the Animal World* (Auckland: David Bateman, 1990).

119 *cats and jealousy* Jerome Neu, "Jealous Thoughts," in *Explaining Emotions*, ed. by Amelie Oksenberg Rorty (Berkeley: University of California Press, 1980), pp. 425–463.

124 *while this behavior has been observed* See his "Social Dynamics, Nursing Coalitions, and Infanticide among Farm Cats, *Felis cattus*," in *Advances in Ethnology* 28 (1987): 1–66.

CHAPTER SIX: FEAR

144 *It has been proposed* "Fear of Animals: What is Prepared," *British Journal of Psychology* 75 (1984): 37–42.

151 *experiments in social isolation of cats* P. F. D. Seitz, "Infantile Experience and Adult Behaviour in Animal Subjects. II. Age of Separation from the Mother and Adult Behaviour in the Cat," *Psychosomatic Medicine* 21 (1959): 353–378.

151 *handling kittens* R. R. Collard: "Fear of Strangers and Play Behavior in Kittens with Varied Social Experience," *Child Development* 38 (1967): 877–891.

152 *Yi-Fu Tuan's* Yi-Fu Tuan, *Landscapes of Fear* (N.Y.: Random House, 1979).

CHAPTER SEVEN: ANGER

159 *dog bites* See Harold B. Weiss et al., in a study supported by the Department of Health & Human Services, reported in the *Journal of the American Medical Association* 279 (1998): 51–53.

160 *Cleveland Amory hit his cat* Cleveland Amory, *The Best Cat Ever* (Boston: Little, Brown & Company, 1993).

166 *Only humans vow to avenge* Richard S. Lazarus and Bernice N. Lazarus, *Passion & Reason: Making Sense of Our Emotions* (N.Y.: Oxford University Press, 1994).

169 *cat's sudden biting* See Bruce Fogle, *The Cat's Mind: Understanding Your Cat's Behaviour* (London: Pelham Books, 1991).

172 *The lights went up* John Bowlby, *Separation: Anxiety and Anger* (vol. 2 of *Attachment and Loss*) (N.Y.: Basic Books, 1973), p. 246.

174 *Human research clearly shows* A summary of the research is found in "Gender and Emotion" by Leslie R. Brody and Judith A. Hall, in *Handbook of Emotions*, ed. by Michael Lewis and Jeannette M. Haviland (N.Y.: The Guilford Press, 1993) pp. 447–460.

175 *social constructionist view* See the collection *The Social Construction of Emotions*, ed. by Rom Harré (Oxford: Basil Blackwell, 1986).

176 *not everyone has the talent for love* Henry T. Finck, *Chopin and Other Musical Essays* (N.Y.: Charles Scribner's Sons, 1889).

176 *this theory suggests* In the work cited in the previous note, pp. 135–166.

CHAPTER EIGHT: CURIOSITY

185 *reflective colors of cats' eyes* For cats' eyes, see Akif Pirincci and Rolf Degen: *Cat Sense: Inside the Feline Mind* (London: Fourth Estate, 1994).

195 *cats and hunting* Z. Y. Kuo, "the Genesis of the Cat's Response to the Rat," *Journal of Comparative Psychology* 11 (1930): 1–30.

197 *Sophia the cat* Personal communication, August 20, 2001, from Pierre Berclaz, the general manager of the American Colony Hotel in Jerusalem.

CHAPTER NINE: PLAYFULNESS

208 *These signals are meant* M. J. West, "Social Play in the Domestic Cat," *American Zoologist* 14 (1974): 427–36. See, too, Paul Martin, "The Four Whys and Wherefores of Play in Cats: A Review of Functional, Evolutionary, Developmental, and Causal Issues," *Play in Animals and Humans*, ed. by Peter K. Smith (London: Basil Blackwell, 1984) pp. 71–94.

211 *irritation of mother cats with kittens' play* Meredith West, "Play in Domestic Kittens," in *The Analysis of Social Interactions*, ed. by Robert B. Cairns (N.Y.: Lawrence Erlbaum, 1979).

212 *Play releases endorphins* According to Jaak Panksepp, the celebrated neuroscientist, there is "widespread release of opioids in the nervous system during play." See his *Affective Neuroscience: The Foundations of Human and Animal Emotions* (N.Y.: Oxford University Press, 1998), p. 293.

224 *kittens' need to be exposed to play* See P. Bateson and M. Young, "Separation from the Mother and the Development of Play in Cats," *Animal Behaviour* 29 (1981): 173–80.

226 *Alexandre Dumas's cat* Kathleen Keto, *The Beast in the Boudoir: Petkeeping in Nineteenth-Century Paris* (Berkeley: University of California Press, 1994) p. 129.

228 *Picasso's* Cat Devouring a Bird See Elisabeth Foucart-Walter and Pierre Rosenberg, *The Painted Cat: The Cat in Western Painting from the Fifteenth to the Twentieth Century* (N.Y.: Rizzoli, 1988).

Recommended Reading

HISTORY

Juliet Clutton-Brock. *Cats Ancient & Modern*. Cambridge: Harvard University Press, 1993.

Donald Engels. *Classical Cats: The Rise & Fall of the Sacred Cat*. New York: Routledge, 1999.

Gillette Grilhe. *The Cat and Man*. New York: G. P. Putnam's Sons, 1974.

Howard Loxton. *99 Lives: Cats in History, Legend and Literature*. London: Duncan Baird, 1998.

Claire Necker. *The Natural History of Cats*. New York: Dell, 1977.

Katherine M. Rogers. *The Cat & the Human Imagination: Feline Images from Bast to Garfield*. Ann Arbor: University of Michigan Press, 1998.

LITERATURE

Joyce Carol Oates and Daniel Halpern. *The Sophisticated Cat: A Gathering of Stories, Poems, and Miscellaneous Writings about Cats*. New York: Dutton, 1992.

Michael J. Rosen, ed. *The Company of Cats: 20 Contemporary Stories of Family Cats*. New York: Gallahad Books, 1992.

BEHAVIOR

Muriel Beadle. *The Cat: A Complete Authoritative Compendium of Information about Domestic Cats*. New York: Simon & Schuster, 1977.

John W. S. Bradshaw. *The Behaviour of the Domestic Cat*. Oxford: C. A. B. International, 1992.

Paul Leyhausen. *Cat Behavior: The Predatory and Social Behavior of Domestic and Wild Cats*, tr. Barbara Tonkin. New York: Garland STPM Press, 1979 (orig. pub. in German, 1956).

Roger Tabor. *The Wild Life of the Domestic Cat*. London: Arrow Books, 1983.

Dennis C. Turner and Patrick Bateson, eds. *The Domestic Cat: The Biology of Its Behaviour*. Cambridge: Cambridge University Press, 1998 (2nd rev. ed., 2000, contains new chapters).

CARE AND ADVICE

Michael W. Fox. *Understanding Your Cat*. New York: Bantam, 1977.

Ingrid Newkirk. *250 Things You Can Do to Make Your Cat Adore You*. New York: Simon & Schuster, 1998.

Roger Tabor. *Understanding Cats*. London: David & Charles, 1995.

David Taylor. *You and Your Cat*. New York: Alfred A. Knopf, 1991.

INDIVIDUAL CATS

Cleveland Amory. *The Cat Who Came for Christmas*. New York: Penguin, 1988.

Roger A. Caras: *The Cats of Thistle Hill: A Mostly Peaceable Kingdom*. New York: Simon & Schuster, 1994.

Doris Lessing. *Particularly Cats*. New York: Alfred A. Knopf, 1991 (orig. pub. 1967; 2000 edition, with the essays "Rufus, the Survivor" and "The Old Age of El Magnifico," Short Hills, N.J.: Burford Books).

Marge Piercy. *Sleeping with Cats*. New York: William Morrow & Co., 2001.

May Sarton. *The Fur Person*. New York: W. W. Norton, 1983 (orig. pub. 1957).

GENERAL

Roger A. Caras. *A Cat Is Watching: A Look at the Way Cats See Us*. New York: Simon & Schuster, 1989.

Paul Gallico. *The Silent Miaow*. New York: Crown, 1964.

Frances and Richard Lockridge. *Cats and People*. New York: Kodansha, 1996 (orig. pub. 1950).

Howard Loxton. *The Noble Cat: Aristocrat of the Animal World*. Auckland: David Bateman, 1990.

Ida M. Mellen. *The Science and the Mystery of the Cat*. New York: Charles Scribner's Sons, 1949.

Fernand Mery. *The Life, History and Magic of the Cat*, tr. from the French by Emma Street. New York: Grosset & Dunlap, 1968.

Desmond Morris. *Cat World: A Feline Encyclopedia*. New York: Penguin, 1997.

Elizabeth Marshall Thomas. *The Tribe of Tiger: Cats and Their Culture*. New York: Simon & Schuster, 1994.

Cal Van Vechten. *The Tiger in the House*. New York: Dorset Press, 1989 (orig. pub. 1952).

Index